I0487095

The Family Guide to Disruptive Climate Change

ALEX COOK

The Family Guide to Disruptive Climate Change

Outskirts Press, Inc.
Denver, Colorado

The opinions expressed in this manuscript are solely the opinions of the author and do not represent the opinions or thoughts of the publisher. The author has represented and warranted full ownership and/or legal right to publish all the materials in this book.

The Family Guide to Disruptive Climate Change
All Rights Reserved.
Copyright © 2010 Alex Cook
v3.0

Cover photograph: The Rocky Mountain Continental Divide at 12,000 ft. in a recent late summer. Photograph by the author from Panorama Point, Golden Gate Canyon State Park.

This book may not be reproduced, transmitted, or stored in whole or in part by any means, including graphic, electronic, or mechanical without the express written consent of the publisher except in the case of brief quotations embodied in critical articles and reviews.

Outskirts Press, Inc.
http://www.outskirtspress.com

ISBN: 978-1-4327-5779-3

Outskirts Press and the "OP" logo are trademarks belonging to Outskirts Press, Inc.

PRINTED IN THE UNITED STATES OF AMERICA

Table of Contents

Acknowledgements

I must first acknowledge the work of those scientists in the field—those glaciologists, oceanographers, meteorologists, and biologists—who have documented the existing climate changes in our backyards and in very remote areas of the world. Their reports have demonstrated the reality of climate change and have supported the analyses of laboratory scientists and the calculations and predictions of climate modelers. I am indebted to the statements and writings of renowned scientists like James Hansen and John Houghton who have made unequivocal declarations of the urgency of the danger of future climate changes. Numerous scientists from the laboratories of NOAA, NASA, and NCAR, as well as the science departments of universities with active programs in atmospheric science similar to those of Harvard, Princeton, Colorado, and Arizona, and the laboratories of England, Germany, and New Zealand have contributed immensely to the vast published literature of the earth's physical, chemical, and biological environment. My modest contributions to research observations in atmospheric science and in communications to students and interested citizens have benefited greatly from the work and writings of those in the above scientific communities.

Finally, the professional editing of Pamela Van Scotter and Leslee Schmitt will undoubtedly make this Family Guide a more readable and valuable resource for understanding the challenges of disruptive climate change.

Preface

Global warming with disruptive climate change is caused by the heat trapping properties of certain atmospheric gases increased by human activities. This is a complex problem with important ramifications affecting our choice of energy sources and our methods of food production. The effects will not be uniform but are nevertheless global; consequently, as with the ozone problem of the late twentieth century, the corrections will require international cooperation.

That global warming is already occurring has become increasingly apparent with improved satellite observations of decreasing summer Arctic ice and worldwide reports of melting glaciers. Migrating birds and insect populations are responding to climate change by inhabiting new regions. Human inhabitants of islands and coastal areas find rising seas encroaching on their homes; dwellers of city and countryside wonder at the connection between global warming and the increased frequency and severity of weather events.

The threat of disruptive climate change hovers over our daily concerns and preoccupations with economic and political issues. Individuals at every level of society need to be aware of the response of the universal natural laws to human activities and to correct the mistakes that jeopardize our future.

The science of climate change due to global warming has matured significantly in recent years with increased global observations and verification of theoretical calculations. Communication of the environmental threat to the general public has lagged considerably and has suffered from political and economic misinformation. I am hopeful that family or classroom discussions of the science might help form a proper foundation for personal and societal actions.

Various feedback comments on the science discussions that I have presented in the preliminary drafts of this Family Guide To Disruptive Climate Change, and in my earlier publication of The Greenhouse Effect-A Legacy under the pen name Alex Cook, have been illuminating and helpful. Evidently the science education background of many readers has not been sufficient to make these discussions an easy first reading. The unfamiliarity of the science vocabulary and exposure to new science laws has required more than casual attention. This has interfered with the final message of the clear and present danger of global warming leading to disruptive climate change. I am hopeful that my presentation of a hypothetical middle school town hall discussion might be a more effective introduction to this urgent matter. The additional organized presentation of the relevant science, adaptations, and solutions is designed to furnish further useful guidance for parents and teachers.

In Section A of the Guide, I present a hypothetical town hall discussion by middle and senior school students of the elementary science of global warming and climate change. Unlike their elders, who might have a limited view of the future, young people realize they are threatened by future climate changes that may adversely affect their lives. While daily news dedication to politics, the economy, and anything sensational occupies the full attention of grownups, young people endure this as everyday static. They are aware on some level, however, of the ominous environmental rumble of starving polar bears and the specter of dying coral. This town hall presentation is predicated on the students' attitude of serious concern for their future. In utilizing this format I am able to introduce some of the confusions and concerns prevalent in today's society and clarify the factual basis for global warming and disruptive climate change.

In Section B of the Guide, I have prepared for parents and teachers an organized review of the important aspects of the problem. I have added several references to easily available scientific reports as well as recommendations for personal response to global warming. The suggested additional readings include a number of interesting and useful discussions by reputable writers on the problems of climate change.

I must insist that attention is required. A continuing failure to read science instructions may cause Mother Earth to take revenge! The reader should understand "Mother Earth" to mean the universal laws of natural science. These laws operate independently of man's aspirations, but human activities sometimes cause natural responses that degrade the quality of life on the planet. Knowledge of the consequences of such actions should permit the implementation of intelligent decisions to attempt to correct these mistakes.

Mistakes in social programs or international affairs can be corrected, but the universal laws of nature are not subject to modification. There is now an urgent need for policy decisions and activities to correct mankind's mistakes in energy use and production that are adversely modifying the climate of the earth. The first step is to educate correctly on the scientific realities of the changes we are facing.

Alex Cook

SECTION A. STUDENT TOWN HALL DISCUSSIONS

INTRODUCTION

A teacher plan for class discussion of climate change might be organized as follows:

Hello. My name is Geraldine Murphy. I teach physical science to the students at Foothills Middle School. The recent media attention to global warming has inspired me to initiate a student project on this subject. I devote an initial class period to make reading assignments, give advice, and offer assistance. Near the end of term we have a town meeting with audience participation from the middle school class in response to a presentation by a panel of senior physics class volunteers. These presentations and town hall discussions may be useful for individual reading assignments or enactment as a science play, or for general family discussions. The following format is a verbatim record of these discussions.

DAY ONE

WE BEGIN WITH THE SCIENCE

Ms. Murphy: Hello Class. As I promised you at the beginning of the term, we are going to have a panel discussion on Climate Change and Global Warming. The panel is composed of senior high school physics students who have expressed interest in the science of global warming. Each senior has been assigned a particular topic to research and report to you today. The presentation will be followed by a town hall discussion with our middle school student body.

This topic has had some attention in the news lately and all kinds of people have been expressing opinions on talk shows and editorial columns and letters to editors.

(Comment from somewhere in the back of the audience) Sure. Right after scandals in Hollywood and Washington.

That's true, partly because most of the media people have little background in science, and they assume their audience doesn't care. But the problem is urgent and we are going to give it the study it deserves. Al Gore's documentary, *An Inconvenient Truth,* will encourage you to get very worried about the problem. And his attention to global warming earned him a share of the Nobel Peace Prize for 2007. But the CEO of a coal company doesn't want you to stop burning coal to generate electricity, and of course an Audubon Society person will tell you that global warming will cause the polar bears to become extinct. So whom do you believe? Many people either don't understand the science or just want to ignore the situation—or both. And most politicians will make laws to keep their voters happy, no matter if they understand the science or not. So, our panel of researchers is going to give us the scientific facts. We have an atmospheric scientist as a guest today to make sure we have it right. Let me introduce Dr. Richard Hawkins, who comes to us from our University down the street. Professor Hawkins--

Hawkins: Good morning to the young ladies and gentlemen of this fine Foothills Middle School audience. And a special greeting to our panel of senior student researchers. My name is Richard Hawkins; I'm an atmospheric scientist volunteering today as moderator for this discussion of global warming.

May I first add to what Ms. Murphy had to say about global warming. We're going to be talking about science. In these discussions of climate change, we will begin with careful observations. Then we will look for explanations. When we find an explanation that fits all the observations, we may call it a theory. Following this, we must look at calculations, using all the correct laboratory measurements to be sure the theory is correct. Finally, we will examine the predictions based on this body of known science. With that information, we will consider the possible responses that we might make to this problem.

Your researchers on this panel have been assigned topics ranging from observations to theories of global warming. These are found in the scientific literature—articles and books that follow the methods I have just described. I believe Ms. Murphy has a list of those references. Your classmates will be telling us what they have learned from their study. You will be able to think about it and ask questions to understand and test the results.

DAY ONE

I. THE SCIENCE OF GLOBAL WARMING
A. *GLOBAL TEMPERATURE*

Hawkins: Let's get started. I expect you may know our panel members for today's discussion. We have Sarah Perkins, Caleb Andrews, Amanda Benson, Mary Anne McIntyre, and Jason Jacobsen.

First we will consider the Global Temperature Record. When someone says global warming, we automatically think of warmer temperatures. Our specialist on global temperature is Sarah, who will lead the panel discussion for about 15 minutes. Then we will break for audience questions for about 10 minutes. Sarah—

Sarah: Thank you Professor Hawkins. We all know about temperature. It's the number on the thing outside the kitchen window—or maybe the thing the nurse sticks under your tongue or in your ear. At different times of the year, the air temperature tells you what clothes to wear. If the temperature is low, you feel cold; your body loses heat. But in summer, the air feels hot and you sweat trying to keep cool.

Hawkins: Excuse me Sarah. I believe Caleb already has a question.

Caleb: But the temperature is just a number. Just what is it that is being measured, like in the air?

Sarah: Well, let's think of the air as being made up of very, very small ping-pong balls. Actually, it's most-

ly nitrogen and oxygen—what we call molecules. The air is mostly just space, with these molecules bouncing around in collisions with each other. They have different speeds and they have energy of motion. We call this heat energy. If the number on the thermometer is high, we say it's hot. That's because the molecules are moving very fast—they have lots of energy of motion. And a low number means it's cold, and the molecules are not moving so fast. That's what the thermometer measures.

Caleb: But there are lots of different thermometers. How can we compare the numbers?

Sarah: We agree to have them give the same reading when water freezes—that's 32°—or boils—that's 212°. That's on the Fahrenheit scale that we are most familiar with in this country. Most of the rest of the world uses the Celsius scale, which makes more sense I think. In the Celsius scale it's 0° for freezing or 100° for boiling.

Caleb: Isn't there also an absolute scale, called Calvin?

Sarah: Yes, there is an absolute scale; the zero on that scale is when the molecules stop moving—there is zero energy of random motion. This happens at minus 273° Celsius. Scientists use it, but it's not very useful for everyday measurements. And it's a scientist's name, Kelvin. I think Calvin has to do with religion.

DAY ONE

So, let me continue. To get an average global temperature, we can have someone at each spot on the earth taking daily readings and then we can calculate an average temperature for that spot for a whole year. Then we can average these averages together to get an average temperature for the entire earth, including the land and the oceans at the equator and the north and south poles. At the present time, that turns out to be about 58° Fahrenheit or between 14° and 15° Celsius. That record of past temperatures was presented by the scientists on the Intergovernmental Panel on Climate Change (that's the IPCC) in 2007. The record shows lots of year-to-year variations, mostly due to weather changes from day-to-day and for different regions. But the graph shows a gradual upward tilt from about 1850 to 2000. If you draw straight line through those points, you get an increase of 0.74° C for the 100 years ending in 2005. And the tilt gets steeper in the most recent years—almost double that rate. That's global warming!

Hawkins: Thank you Sarah. That was a fine presentation. You covered the topic nicely. Now let us see what questions the audience has.

(Audience) I'm Sean Baxter. I keep hearing different numbers for that temperature increase. What is that in Fahrenheit?

Sarah: On the Fahrenheit scale there are 180 divisions between freezing and boiling; on the Celsius scale there are 100 divisions. So a change of 1° C would be 1.8° F. And 0.74° C is 1.33° F.

Hawkins: Another question:

(Audience) I'm Whitney Jameson. So global temperature changed from 57° to 58°. That's not much of a change. I probably wouldn't notice that.

Sarah: But some place it has changed from 32° to 33°. That means that ice melts. That makes a big difference to tulips and chipmunks. Winters get a little bit shorter—and summers get a little hotter.

(Audience) Whitney: We think we are lots smarter than chipmunks; somebody ought to be able to fix that temperature. Can't we just let the government do something about it?

Sarah: Well, first we need to understand it, like we are trying to do right now. With global warming, the climate will change in lots of ways. Human beings have adapted to the land and the oceans and the weather that we have already experienced. Things could get very difficult when the climate changes. The IPCC says that we are part of the problem; we all need to work to fix it.

(Audience) Sean: This IPCC report. Where can I find it?

Sarah: It's on the Internet. Just Google IPCC. Then click on Climate Change 2007. It's a summary for policymakers, and it's only about 20 pages long.

DAY ONE

B. *The Greenhouse Effect*

Hawkins: So, we've got global warming. But what causes it? Caleb---

Caleb: First, I have to tell you about the greenhouse effect. Maybe you know about greenhouses, where people grow flowers or vegetables when the outside conditions are too cold. Most have glass roofs so the sunshine gets in. And after a while the air and ground inside gets warm enough for the plants to grow. Why does this happen?

Hawkins: I think Amanda is eager to tell you.

Amanda: That sunshine carries a lot of energy from the sun. Some of it gets absorbed by the plants to make them grow, and some is changed into heat energy to keep them warm. And if it gets too hot, we can let some of the hot air escape by opening the windows.

Caleb: Right! That's part of the story. Like you say, that sunlight comes through the glass. Let's think about that. You've seen rainbows. That's the visible sunlight separated into its spectrum from violet to blue to green to yellow to orange to red. That's as much of the sunlight that we can see. But there is more radiation out there that's not so obvious.

Amanda: I know. There is ultraviolet and infrared too. We just can't see it.

Caleb: Yes, you've heard about ultraviolet, beyond the violet. We can't see it, but it will give you a sunburn if you get too much—it's called UV-B.

Amanda: But isn't the UV-B absorbed by the glass so it can't give us a sunburn or heat the greenhouse?

Caleb: That's right. Actually, most of the sun's ultraviolet is absorbed by ozone in the stratosphere; the ozone layer partially protects us from sunburn. So ultraviolet is not the really important cause of global warming.

(Audience) My name's Jerry Smith. Hey, you just lost me. You said stratosphere. I know it's up high somewhere, but where? And why is it called that? And what does it have to do with global warming?

Hawkins: The stratosphere is the layer of the atmosphere just above where most commercial jets fly—about 35 to 40,000 feet, 7 or 8 miles above sea level. The lower atmosphere down to the earth's surface where we live is called the troposphere. This is where we have weather with clouds and precipitation. The air in the stratosphere is very stable—there is no convection to make clouds. The climate is what happens in the entire atmosphere but what we see is mostly what happens in the troposphere. So, Caleb—why do we have global warming?

Caleb: There is also infrared, beyond the red. We can't see that part of the sun's spectrum either, but scientists can detect it with instruments. And they find that it too is absorbed by the glass. So what is infrared, sometimes just called IR? Well, for example, if you heat an iron poker in the fireplace or with a torch, it will begin to glow red. We say it is red hot. And if we heat it enough, it will look white—call it white hot.

Amanda: But if you let it cool, the colors disappear; it looks normal black again.

Caleb: If you move your hand near the poker when it's red, you can feel the heat. And as the red color fades and it begins to look normal, you can still feel the heat—you know not to touch it. That heat radiation that you can feel but can't see is the infrared. Whether the poker looks white or red or black depends on the temperature.

Sunlight is white—the sun's surface is white hot! The ultraviolet and the infrared are always there in the sunlight, but the visible light is brighter. The sun's spectrum of radiation increases in intensity from the ultraviolet to the visible, then decreases gradually all the way out into the far infrared. The spectrum of heat radiation has the same general intensity change from the ultraviolet to visible to infrared for all materials and depends only on the temperature. All that radiation carries energy; the temperature of the material determines which part of the spectrum is brightest, and the total amount of energy radiated. For objects with temperatures lower than that of the sun, the amount of heat is less, of course, and the strongest part is in the infrared.

Amanda: But it's just the visible light that heats the greenhouse. Isn't that all that's important in the greenhouse effect?

Caleb: Oh, no! The dirt in the greenhouse is warmed by the visible sunlight and it may look normal like the poker, but it is radiating heat energy in the infrared. It's not hot enough to glow red of course. And that infrared can't get back out through the glass. That infrared heat energy is trapped inside the greenhouse. So how warm the greenhouse gets depends on the amount of visible light energy coming in and the amount of the infrared heat energy that is trapped. That is what we call the greenhouse effect.

Hawkins: And now I'll bet that Mary Anne is going to explain what the greenhouse effect has to do with the earth's temperature.

Mary Anne: Right! Now it's obvious that the bright white radiation from the sun is carrying energy that heats the earth when it is absorbed. In fact, NASA has a satellite measuring the amount of heat that arrives at the top of the atmosphere. It's 2 calories per minute, per square centimeter. Those units probably don't mean much to you, but if you could trap all that heat on the roof of your house to heat your bath water, it would be too hot! And scientists have learned enough about the physics of infrared heat radiation to calculate how warm the earth's surface should be—a global average temperature. The visible radiation coming in and the infrared radiation going out have to balance. The earth can't control the temperature by opening windows like in a greenhouse. Dr. Hawkins suggested that it would be easy to find the answer for a simplified case of an earth with no atmosphere. He showed me a physics formula that will give the answer. I did it on my laptop. The answer was 7° Celsius.

Amanda: But the scientists' measurement was more than 14° C. You got the wrong answer. And some of that sunlight is reflected from the ground back out to space. Did you allow for that?

Mary Anne: If we make a correction for that, it makes the answer even worse—that would give minus 7° C. That's below freezing. And we haven't even considered reflection from clouds; that's more complicated. Those temperatures would not permit life as we know it on our earth.

Hawkins: Let's see if the audience understands what was wrong with that calculation.

(Audience) I'm Evan Daniels. It's obvious to me that you got the wrong answer. And you didn't say anything about the greenhouse effect. But what do greenhouses have to do with it? The earth doesn't have a glass roof.

Mary Anne: Of course not! But it has an atmosphere.

(Audience) Evan: But the air is transparent—except for clouds.

Mary Anne: That's correct—for visible light. And, yes, clouds are important—and we have lots of water. Now you know that when water evaporates, it just disappears into the air. Then these molecules of water, that we call water vapor, share the energy of motion just like the nitrogen and oxygen, and like those other molecules, they don't absorb visible light. But scientists learned long ago that water molecules do absorb strongly at certain places in the infrared. They behave like the glass in the greenhouse. They absorb much of the infrared heat radiation from the earth's surface. That's the greenhouse effect. These and other molecules like carbon dioxide and methane and nitrous oxide—we call them greenhouse gases—act like a nice warm blanket and keep the earth about 21° C warmer than if there were no atmosphere.

(Audience) Evan: So the greenhouse effect is good.

Mary Anne: Yes, but we can have too much of a good thing. A thicker blanket of greenhouse gases will produce additional global warming. I think Jason is about to explain.

Jason: People have known about this nice atmospheric blanket for a long time. And we've known that burning coal and oil puts extra carbon dioxide—that's CO_2—into the air. There has always been a concern that the increase of this greenhouse gas would change the balance of radiation in—radiation out. And scientists realized that we were unintentionally doing an experiment to find out! The earth's population was increasing and we were burning a lot more coal and oil in industrial activity. The carbon from those fossil fuels was changed to CO_2 in the air by the oxidation— that's just a chemical word for burning.

So about 50 years ago, scientists began taking measurements of the CO_2 in the air above Mauna Loa in Hawaii. That's a high mountain, so variations like morning traffic or local industries won't affect the results. And there were some calculations that indicated that if we doubled the amount of CO_2, the earth's average temperature would increase several degrees Celsius. At first, this was just a kind of classroom exercise—not something to worry about right away. But the global measurements of temperature and CO_2 were becoming important. And now the IPCC, The International Panel on Climate Change, says that the measurements on Mauna Loa compared to the air trapped in the Greenland ice show that the CO_2 has increased by about one-third, from 280 ppm before the industrial revolution back in the 1700s, to 379 ppm in 2005.

Mary Anne: I don't understand. What are those numbers?

Jason: ppm stands for parts per million—the number of CO_2 molecules you find in a million molecules of air. The IPCC says that is now the highest concentration of carbon dioxide in 650,000 years! The greenhouse effect of CO_2 is increasing rapidly and the temperature record is showing global warming as the response.

Hawkins: Looks like you have a comment from the audience.

(Audience) George Manning here. My Dad has a job at a place that makes cement. He says that when they heat the limestone to get cement, a lot of CO_2 goes into the air. And I guess burning the fuel to get heat makes a lot more.

Hawkins: That's true. That makes a lot of carbon dioxide contributing to the atmospheric increase, and it's a problem that won't be easy to fix. Still, I think it's possible; it would just cost more to manufacture cement. And that's the basic problem; we've been doing things the cheap way without paying for the environmental damage of increasing the greenhouse gas. OK. Back to the audience again.

(Audience) I'm Jeremy Sorenson. Some people say that CO_2 is not a pollutant—doesn't hurt to breathe that little bit. In fact we breathe it out as we breathe in oxygen. And it's good for the plants, right?

Jason: Yes there has been some carbon dioxide in the air for a very long time. When the plants showed up hundreds of millions of years ago, they used some of the sun's energy to grow, and changed some of the CO_2 into O_2 in that process. As more O_2 became part of the environment, this condition allowed organisms that used the O_2 for their life processes to evolve. And when the plants died, their carbon molecules got buried and changed into coal and oil. Our problems all began when our ancestors discovered fire. Burning almost anything changes carbon compounds into carbon dioxide and that's a greenhouse gas. At first it wasn't a problem—there were just a few of us wandering around gathering fuel and seeds and hunting other animals. We thought we were conquering nature. But now there are lots of us, and the truth is, we are part of nature. There are so many humans that we are burning a lot of that coal and oil to make energy for our houses and factories and cars and trains and airplanes. OK guys. Here's the bad part. If we make that greenhouse effect blanket thicker, we'll wake up sweating. And we ARE making it thicker.

Hawkins: You've given me an idea. I should write a book called "My Life" by Car Bon -- the story of a powerful atom. Of course there would be the usual warning: Readers should use Caution—contains educational material. Adults may be exposed to science and become frightened.

Jason: Ah, but you're just a scientist. It would never get published. You need to have a reputation like J.K. Rowling, the author of the Harry Potter books.

Hawkins: Well, I could write it as a children's book too. Like: I was locked up for a million and one years with my brothers and sisters under tons of dirt. I was rescued from that prison by the oxygen twins and we floated away in the air, free as birds. And when the red photons joined our party, we began to dance. Those skinny little fellows had lots of energy, and we trapped them so they wouldn't leave the party.

Mary Anne: I'll bet you had a really hot time!!

Hawkins: Then I was forced to work as a slave for some simpleminded two-legged creatures. They didn't understand the rules of our games and got into all kinds of trouble.

Jason: I know: Some of our politicians are pretty ignorant of those rules. Maybe Bill Gates would sponsor a project to educate them.

Hawkins: Sorry for the distraction. Back to questions on climate change from the audience.

(Audience) Jeremy: Why is climate change with global warming a problem all of a sudden? I hear people complain that we can't afford to change energy production just now.

Jason: You're right. That's one of the arguments against making new rules on greenhouse gases. But other countries, like Germany and Japan, have been paying attention to the science and are way ahead of us on solutions.

Hawkins: I've got an example of a car problem that might help you to appreciate the need to work on slowing greenhouse gas emissions right away. Suppose you are riding in your car on the Interstate and the red oil pressure light comes on. Your Mom says, "Better pull over. I'll call Triple A." But your Dad wonders, "I think there's something wrong with the detector in the engine. We'll be late for the movie—maybe it can wait until we get to that exit." And Mom objects, "That red light means something is really wrong! If we don't stop now, it will cost a lot more to fix it."

You see, the science of global warming is correct and reliable. We now understand the science well enough to know that if we don't decrease the greenhouse gas emissions soon, the costs of adaptation will be a lot more.

(Audience) Jeremy: But the climate has always been changing, like the ice ages; the glaciers are still melting from that. What about natural variation? Or maybe the sun's radiation is changing a little.

Jason: Yes, and our scientists are gradually learning more about the reasons for those past changes. I've heard explanations for a couple of them. About the time the last big ice age ended, the earth wobbled just a little in its orbit because of gravity from the other planets, and the tilt of the earth's axis increased the amount of radiation on the northern hemisphere. So the temperature increased just a bit and the ice began to melt. Then there wasn't as much reflection of sunlight on ice and it warmed a bit faster. Finally, the temperature reached the tipping point for the CO_2 from the oceans to go into the atmosphere and make a sudden large increase in the greenhouse effect. It got a lot warmer.

And in 1600 to 1800 during what was called the little ice age, there were no sunspots. This probably means there was less solar radiation at that time. But the earth's orbit doesn't change very fast and is stable just now, and we have satellites measuring the solar radiation so we know that the changes now are not enough to cause the global warming that we observe. The report by the scientists of the IPCC concluded that the global warming is real and that it is very likely caused by human activity. The increased greenhouse effect of CO_2 and a few other pollutants is responsible.

(Audience) I'm Don Larson. But why blame it on carbon dioxide? There's a lot more water than there is carbon dioxide.

Jason: Well, nature has been working with that for a very long time and is very efficient at maintaining the balance. Water evaporates from the lakes and oceans and the water vapor works for the greenhouse effect. But then it rains or snows in a week or so, taking the water back out of the air, and the cycle starts over. The amount of water in the air keeps changing with the days and the seasons, but the water cycle is only about 10 days and it doesn't have a chance to increase without limit. The amount of water vapor is just right for life as we know it. We know of no scientific reason it should change, if we don't mess it up.

Hawkins: Ay! But there's the rub! If we let the earth get warmer by burning fossil fuels, we may get a nasty surprise if the water responds with a big positive feedback.

For example, as the oceans get warmer, there is more evaporation that puts more water vapor in the atmosphere; that's a big increase in the greenhouse effect.

Caleb: But there's a carbon cycle too. We burn coal that puts CO_2 in the air and the trees take it out.

Jason: Yes, but the carbon cycle is much longer. Things burn and put CO_2 into the air and plants take it out to grow. Trees growing and decaying take a century or so. And some of the CO_2 goes into the ground and the ocean, which means that the carbon cycle in the environment is a hundred years or more and we can't easily correct our mistakes and go back to square one. We have been putting stuff in the air faster than the trees can take it out and we're making things worse by cutting a lot of the trees. We're part of nature, and if we aren't smart enough to get things back into balance, the final result could be a disaster.

Hawkins: Let's stop and see if we really understand the problem. Are there questions from the audience?

(Audience) My name is Alicia Swenson. I'm not sure that I understand. It's the extra heat energy that we put into the air from burning fossil fuels?

Mary Anne: It's true that we are adding heat energy to the atmosphere, but it is thousands of times less than the heat energy we get from the sun. The heat energy that we are adding to the atmosphere is not the prob-

lem. It's the increased greenhouse effect of the CO_2 that's trapping more of the infrared heat that's bad.

Hawkins: Yes, it's the increase of greenhouse gases that is causing global warming. Maybe a familiar example will help us all understand better. I suspect that most of you are thinking about getting a license and learning to drive a car. And I am sure that your parents are concerned about having you drive safely and not have accidents. As a driver, you have control over a big powerful machine. You do it with just the steering wheel, gas pedal, and brake. And, if you don't use your brains, if you drive recklessly or fast, there is a high probability that you will have a serious accident.

In a similar way, the earth's climate is a powerful machine using a very large amount of solar energy. It's the balance of radiation that determines the temperature. And that's controlled by the amount of greenhouse gases.

(Audience) I'm Rosa Barker. I get it! The amount of greenhouse gas in the atmosphere is like the brake and gas pedal.

Hawkins: Very good! And we're the drivers because we have our foot on the CO_2. And if this big machine is going too fast around a curve there's going to be a wreck!

(Audience) Rosa: And no one in the car survives because it's global.

Hawkins: You got it! And if the car is going too fast to be quickly stopped by the brakes, then you won't have time to avoid a crash.

DAY ONE

Amanda: That's like having too much CO_2 in the atmosphere; we can't just take it out of the atmosphere to slow down the car before we go over the cliff.

(Audience): Alicia: But sometimes there's somebody in the back seat who wants to go faster because it's more fun. That puts a lot of pressure on the driver.

Hawkins: I'm sure that's true. It's the responsibility of the driver to drive safely. We're all riding along on the planet; some of us are using more energy than others and making more CO_2. We're all part of the problem. We can't just blame it on a driver who maybe doesn't even speak our language.

C. ENERGY TRANSPORT

Hawkins: You know, in the greenhouse down the street they can open windows to let excess heat escape and they can also turn on fans to circulate the air and warm the cooler corners of the building. We know that the earth doesn't have the option of transporting heat out into space except by radiation, but it can transport energy from one part of the earth to another. That's why the IPCC report says that there will be global warming effects on the weather. I think that we need to look at some geography to understand this. Emily--

Emily: Right! This is important; pay attention! On the earth, we have day and night and we have the seasons. Day and night is pretty obvious. The earth spins on

an axis through the north and south poles one complete revolution every 24 hours. We have day when our side of the earth faces the sun, then it's night when it doesn't—over and over again. And the earth is moving through space; the sun's gravity makes the earth move in an almost circular orbit with the sun at the center. Global warming won't affect either of these.

(Audience) Jacob: Oh, goody!

Hawkins: Glad that makes you happy. Would you like to continue with our geography lesson, Emily?

Emily: Sure. The earth is round--maybe you knew that? Now think how the sun shines on this round ball. The part of the earth that faces the sun directly—we call it the tropics—gets the energy squarely on each square foot—or square meter if you've learned to think in metric units—of the surface. But if you look a little way to the north or south, that same amount of energy is spread over a larger surface area. So the amount of heat generated on that surface area is less. And eventually you get to a place where the sun's rays just graze the surface—here there is really no heat from the sun. So you would expect the tropics to get extremely hot, and the polar regions to be extremely cold. And in fact, I've read someplace that's exactly what happens on the moon. But on the earth, we have oceans and an atmosphere that move the heat energy from the hot places to the cold places. This moderates the extremes that would exist otherwise.

DAY ONE

Mary Anne: So how can you transport heat? It's not like you can put it in a bucket and carry it to some-place cold, is it?

Emily: Now you're being silly. No, that won't work. But in the ocean the Gulf Stream is a current of warm water moving from the Caribbean and Gulf of Mexico across the North Atlantic towards Europe. That warm water and the atmosphere above it makes the average annual temperature in London about the same as Boston, which is 600 miles further south. But how does this work? The earth is a spherical ball; the force of gravity is just about the same over all the oceans. But the ocean water isn't all the same density. For example, the Gulf Stream water loses heat as it moves into the cold polar region and the molecules stick more closely together. And some of the water evaporates and it becomes more salty, so it is more dense and sinks to the bottom. That cold salt water can make a return current back to the south on the ocean bottom. The warm water at the top flowing north and the cold saltier water on the bottom flowing south trans-port about one-third of the heat energy from the tropics to the Arctic. The climate of Europe is mild and pleas-ant because of the warm surface water from the tropics.

Mary Anne: So I guess the warm air moving north from the tropics does the rest? And how does this make weather like blizzards and thunderstorms? And what about the seasons?

Emily: Yes, the seasons are a complication. Let's look at that. Just imagine the sun is there on the floor at the center of our circle of chairs and the earth's orbit is on the floor right here at our feet with the northern hemisphere on top. But the earth's axis of rotation through the north and south poles is not straight up towards the ceiling; it's always pointed at Polaris, the North Star, 23 degrees away from the vertical, a little bit toward the audience. So when the earth is in its orbit there in front of Justin at the back of the circle, the sun is shining squarely on the northern hemisphere. The northern hemisphere where we are gets extra energy: it's summer. And six months later when the earth has moved to the opposite side of the circle, the sun shines squarely on the southern hemisphere. It's their summer, but our winter. And all the time the oceans and the atmosphere are trying to transport the heat from the hot regions to the colder regions.

Mary Anne: OK. I guess global warming makes that a more complicated process.

Emily: You bet it does! The seasons don't have much effect in the tropics. At the equator, the midday sun is always shining nearly squarely on the surface and the days are always close to 12 hours long. Over the land, the daytime temperatures are very high. And the temperature over the ocean regions changes only a little with day and night or with seasons. It takes a lot of energy to change the temperature of that volume of water. The ocean temperature will increase with global warming, but very slowly.

(Audience) Alicia: So nothing happens.

DAY ONE

Emily: Let me finish. In the Arctic, in midwinter everything north of the Arctic Circle never gets any sunlight. That's about 23 degrees change in latitude from the North Pole because of the earth's tilt. When that happens the Arctic loses heat continually through radiation from the surface. The land and the atmosphere get very cold. Usually all the surface water freezes; it changes to ice. The air is cold and dense and the atmospheric pressure builds up until that cold dome of air moves south and causes winter weather in mid-latitudes. Especially in springtime, when the Arctic air meets that tropical moist air, say in Mississippi, the condensation releases a lot of energy and the storms can be very violent.

But in midsummer, that Arctic region gets sunlight continually. Day and night temperatures both increase. Normally we expect much of the surface ice on land to melt, but underground ice—the permafrost—should remain frozen. The Arctic Ocean has always had more open water in summer and the ice always begins to melt on the glaciers and ice cap of Greenland. The trouble is, with global warming more of that ice will melt.

(Audience) Rosa: But if the Arctic is getting warmer, won't that change the energy transport from the tropics?

Hawkins: That's a good point. We've got some original thinking out there! Yes, we're in uncharted territory with climate change. The circulation of the atmosphere and even the ocean may change. But just remember, when the tropical oceans warm, there will be more water vapor in the air—and with more precipitation there will be more active weather. We just can't say exactly where.

(Audience) Alicia: We don't have an ice cap and permafrost here, so maybe climate change won't be so disruptive here in the middle of the country.

Emily: True. In mid-latitudes, like where we are, the changes are not so obvious. The winters are a little shorter and the summers are a little warmer. But those are very large energy changes in the Arctic and they will eventually affect us. Like Dr. Hawkins says, we will begin to see more violent storms because of the additional transport of heat and water vapor from the tropics. And people along the seacoast and on islands will see the sea level rise because of the melt water from the glaciers and the ice caps, and from the warming ocean. Water expands when it is heated, just like almost everything else.

Mary Anne: Wow! That could get to be a problem!

(Audience) Oooooh!

DAY TWO

II. OBSERVATIONS OF CLIMATE CHANGE

Hawkins: OK. We've got global warming and we are the cause. But what are the climate changes? On our panel today, we have Kyle Johnston, Cody Jefferson, Justin Kennedy, Brandon Schwartz, Robert Peeks and Emily Miller. I believe Kyle is ready to start.

Kyle: Well first, weather is what we see changing every day or so. Climate is different; it's the average behavior over a long period of time. Climate is what we have adapted to in our lives. A single windstorm or flood doesn't make the climate, but a changing pattern toward more frequent events like that is a climate change. So let's look at how weather happens and how global warming can change it.

The temperature is getting higher. So we wear different clothes and adjust the air conditioning in our house. That's easy. And if the humidity is higher, we get more thunderstorms with rain, and we have to mow the lawn more often. If the air is dry, the grass dies and the ground gets hard and cracks open. But it's the average temperature that changes the climate. The average temperature depends on where we live, and it changes with the seasons. To see the effects of climate change, we have to look at all these natural responses as the average global temperature increases.

A. *MELTING GLACIERS AND OCEAN ICE*

Hawkins: We've heard the temperature evidence from Sarah; there's other evidence that isn't as obvious because it isn't in our back yards.

Kyle: Yes. Thank you Dr. Hawkins. Sarah explained that a temperature change from 57° F to 58° F wasn't something that we would notice, but increasing from 32° to 33° melts the snow and ice. We've all seen pictures of glaciers that form in high mountain valleys from heavy winter snows. The weight of all that snow and ice makes the glaciers move slowly, only inches a year, to lower altitudes where it melts in summer, or falls into the ocean. But if the air temperature increases to 33° just a few days earlier every spring along with warmer summers, the lower part will melt faster than the top part grows when it snows. And we're now seeing new pictures where the lower part of the glacier, like those in Glacier National Park and in the Andes in South America and in the European Alps, has completely disappeared. It has all melted and the ice foot of the glacier is way up the valley.

Cody: I visited Glacier National Park last summer; the bottoms of the glaciers were gone and the park rangers there say that in just a few years the glaciers will be completely gone.

Hawkins: Looks like a comment from the audience--

DAY TWO

(Audience) Alicia again. So maybe there is something special happening in Montana; that doesn't mean it's global.

Kyle: We can see pictures in National Geographic where almost all the other glaciers in Alaska and South America and Greenland are doing the same thing. And there are government and university scientists who are making measurements at some of those places and reporting the results in scientific journals.

(Audience) Matthew Jensen here. My grandparents took a cruise to Alaska last summer. They saw ice falling from the glaciers into the sea and they took pictures from an airplane of some glaciers that had melted way up in the valleys. But things haven't changed at our house.

Kyle: Things have not changed that much. We always see the winter snow melt every spring. Nothing unusual, except that it is happening just a little earlier, and the first snow sometimes happens just a little later in the fall. Maybe the winters will get shorter and not be so cold. We might like that; we can put the winter clothes away early and enjoy the change. But in a few years, the robins and bluebirds may arrive from their winter homes too late to find their favorite worms or bugs. And some of the flowers may bloom too early and the hummingbirds won't find nectar when they arrive. And if the hummingbirds aren't around to move the pollen from flower to flower at the right time, there won't be any seeds to grow new plants.

Emily: Well, I sort of like the idea of spring coming a bit earlier—it's easy enough for me to follow the change. Won't the birds and animals learn to change the timing of their migration?

Kyle: Sure, to us the change is hardly noticeable, and to some extent it is change that we can adapt to. But some of the other creatures can't adapt so quickly, and many plants and animal populations may die out before their species can adapt. This could soon become a serious problem.

Hawkins: I believe we are ready to hear about some other lines of evidence that are obvious results of global warming. Would you like to continue with that, Brandon?

Brandon: OK. Yes, let's move on. The scientists who are making observations in the Arctic have found that the sea ice is thinner. It's only a few feet thick and it covers less of the Arctic Ocean every summer. I've found pictures taken from the NASA satellites that show the area of white frozen ice is getting much smaller. The Arctic sea ice has thinned and in summer is melting into open water. In 2007, the ice was only about half the amount it was 50 years ago. We don't see these changes caused by the melting ice right before our very eyes, but have lots of pictures. And the Arctic natives—the Eskimo and other tribes—see their shoreline eroded by the open water and some of their villages are falling into the sea. The polar bears can no longer move about on the ice to search for seal meat. And the walrus can no longer rest on the ice between dives for shellfish on the sea bottom.

Hawkins: I think we have an objection from the audience-

(Audience) Matthew: Maybe things like that just happen in the Arctic naturally. But nothing like that is happening here.

(Audience): My name is Janet Martin. I'm not so sure about that. I have a mystery for you. My Aunt Marjory up in the high country feeds the birds in winter. She said the rosy finches have disappeared—there used to be hundreds when the blizzards started.

(Audience): Sam Jenkins here. My Uncle George and Aunt Mary live in a place up along the Peak-to-Peak highway. They invited us up for a barbeque last August and we went for a hike up to the Continental Divide. There wasn't any snow up there! But I remember doing that a year or so ago and saw lots of snow that hadn't melted all summer. Must be global warming, huh?

Emily: My Mom is a member of the Audubon Society. She says the birds don't nest up there any more because the snowfields are gone. The parent birds used to collect seeds and insects from the surface of the snow to feed the baby birds. But maybe they will come back if we have lots of snow again.

Hawkins: Yes, if we get cold winters with lots of snow, things could go back to normal. Weather changes like that have happened often in the past. But if this is a continuing trend, normal conditions won't return in our lifetime.

(Audience): Sam again. Another thing! We always drive over Trail Ridge Road in Rocky Mountain National Park in late summer before school starts. We didn't see any snow this time either. And on the other side of the Divide near Winter Park the trees were all dead!

Hawkins: Yes, the lodgepole pines have been hit hard by pine bark beetles. The Forest Service says the insects are moving to higher altitudes because the winters aren't as cold as they used to be, and it's been dry so the trees weren't healthy so they are more vulnerable. Now they are worried that lightning might start bad wildfires in those dead trees.

Emily: If it's really global warming, won't it keep getting worse?

Hawkins: Maybe, but remember, the IPCC temperatures showed lots of variations. If it continues for several years though, it could well be climate change due to global warming likely caused by human activities.

Brandon: Well, the Arctic climate behavior is a lot different from here and the changes can be even greater there. You probably know that there is no sunshine above the Arctic Circle in midwinter, but it shines 24 hours a day in midsummer. The radiation goes from one extreme to the other. It's always on the border of ice or water. If that changes just a little, they will have a very different climate. The scientists who have been monitoring this situation closely and calculating the global warming effect are very concerned that this region will be ice-free

in summer sometime in the near future. You see, in summer, the area of Arctic Ocean ice gets smaller as the ice melts faster because there is less reflection of solar energy and the radiation balance makes it even warmer—that's called positive feedback. Scientists see no way to avoid an ice-free summer Arctic Ocean. And the satellite measurements of the Greenland ice cap show that it is less thick too, and the edges of the ice cap are disappearing.

Hawkins: And that Greenland meltwater is contributing to sea level rise and may affect the Gulf Stream circulation. Oh, sorry to interrupt; go ahead.

Brandon: So the melting ice is clear evidence of climate change with global warming. We know from our experiment in the laboratory—remember that it takes almost as much heat to melt ice as it does to heat water to boiling—there is a very large amount of energy involved in melting ice. And the temperature of the ice doesn't even change as it melts. So the melting ice in the glaciers and the Arctic is an even stronger indicator of climate change than that few degrees of temperature change everywhere else.

Justin: You know, that sounds a little like an example of the Gaia hypothesis that I've read about. That says that the earth is like a living thing able to protect itself. So the melting ice acts to delay the temperature increase. Eventually we'll correct the cause of global warming and things can go back to normal. Maybe that's nature's way of keeping the planet healthy.

Brandon: But even if we stop putting extra green-house gases in the atmosphere things won't go back to normal quickly—not even in our lifetime. The CO_2 that we put up there yesterday will keep trapping extra heat. Many of those molecules won't get trapped in the ocean or the forests for something like a hundred years and some of them will be in the atmosphere for a very very long time! So even if we correct our mistake immediately we will continue to see climate change. A few of earth's species are still likely to go extinct. And a lot more will go extinct if we wait to fix things until all the ice melts. Besides, we humans may also find it very difficult to adapt to the climate change. We've become an important part of Nature: we're supposed to be smart enough to fix the global warming problem before it's too late.

B. SEA LEVEL RISE

Hawkins: OK. So the warm ocean water is expanding and the fresh water and ice from Greenland is going into the ocean. That means the ocean surface is a bit higher. But the ships still float on the top. So how can that affect the way each of us live? Justin has been reading about the sea level problem. Justin---

Justin: Thanks, Dr. Hawkins. The sea levels have been rising for a long time. Not much: just 17 cm in the 20th century. That's not quite 7 inches. Those measurements come from tide gauges in harbors, and since 1993, from satellites. From 1961 to 2003, the average rate was 1.8 millimeters per year, but the rate increased after 1993 to 3.1 millimeters per year, almost double. (2.5 millimeters is a tenth of an inch.)

Emily reminded us that most things expand when they are heated. All the molecules bounce around more and get further apart. About half of that sea level increase is just from expansion. And if you pour more water in the ocean from Greenland or other areas, the top surface gets higher. The extra water starts creeping up the beach in Florida. Then with any kind of storm in the Atlantic, the sand washes away. So the Atlantic is knocking on the door, and when the people living in homes along the edge of the Florida Everglades find their drinking water is getting salty, the ocean is already in the house and it isn't ever going to go away!

(Audience) Hi, I'm Jeremy Sorenson. You mean, like never? How come?

Justin: A sea level rise and the increased pressure will begin to push salt water into the fresh water aquifers and the wells. The problem was caused by global warming, but the earth can't cool down with that extra CO_2 in the atmosphere. Those greenhouse gas molecules won't go away for at least a hundred years! So Florida tourists may have to learn to drink salt water.

Brandon: So they spend some more money and build desalinization plants—that will fix the problem.

Justin: Sure, but you know if the ocean water gets into the Florida condo parking garages and some millionaire's back yard, those people are going to want FEMA to come save their expensive property. We'll probably start hearing about plans for a levee around all of South Florida—that would take a lot of your tax money. That might protect some of that expensive property for a couple of years—but not if there are more category 5 hurricanes.

It's important to realize that the salt water will be in the houses in Bangladesh too, and the Pacific islands like Tuvalu, and the Maldives in the Indian Ocean will be under water. Those people would like someone to build levees to keep the ocean out, but those governments have no money for projects like that. Actually, on Tuvalu the ocean water just bubbles up through holes in the rocks, so levees won't stop it. Those people will have to move someplace else—maybe to areas where it's already crowded. Talk about an immigration problem—that will be a real big one.

Hawkins: Yes, it's going to be the poor people who will suffer the most from global warming, even though they did little to cause it. That hardly seems fair, does it? Now let's get back to the heat transport from the equator to the poles.

DAY TWO

C. WEATHER

Hawkins: Let's continue our discussion of Climate Change and Global Warming. Weather is what we see changing every day or so. Climate is the average behavior over a long period. That is what we have adapted to in our lives. A single windstorm or flood doesn't make the climate, but a changing pattern of these events is a climate change. Much of the weather occurs because the earth is heated unevenly by the sun. The solar radiation strikes the surface in the tropics at near normal incidence; that is, the sun's rays are perpendicular to the surface. At higher latitudes, the rays are more nearly grazing; the energy is spread over a greater area and the heating is less. Consequently, the earth's oceans and atmosphere act to transport heat energy from the tropics to higher latitudes. There is an additional complication of the seasons: the earth's axis of rotation tilts back and forth from the sun as the earth moves in its orbit about the sun.

We've talked about how the earth's temperature is determined by the balance of radiation in from the sun and out from the earth. We can't open the windows to space to control the temperature, but the earth's atmosphere and oceans can move the heat energy from the tropics to polar regions. And global warming changes that activity. So, we are going to give some special attention to the generation of violent storms.

And we already have a question from the audience.

(Audience) Alicia: Didn't Emily say that the ocean currents transport about one-third of the heat from the tropics? How does that affect the weather?

Brandon: Yes, the ocean transports some of the heat from tropics to the poles. But it's the atmosphere that is responsible for most of our weather. And global warming can change that too, as well as cause sea level rise. That's part of the story. For one thing, moving the warm tropical air molecules to the Arctic releases heat energy when those air molecules cool. But it's the cycle of water in the atmosphere that does the exciting part.

(Audience) My name's Whitney. So, you transport water that way. But what does that have to do with heat?

Brandon: Well, the heat that evaporates the water is stored as energy in the water vapor. When the water vapor condenses, you get it all back—conservation of energy! So energy is taken out of the tropics through evaporation and is moved, often further north, to a lower temperature where condensation releases the energy.

(Audience) Whitney: Then where does it go?

Brandon: Well, some of it goes to increase the temperature—that's the random energy of motion that Sarah talked about the other day. And, of course the wind blows!

(Audience) Whitney: So it's that water cycle that transports the heat?

DAY TWO

Brandon: Right! See, the warm ocean water in the tropics is slowly evaporating. After a week or so, there is a lot of water vapor in the atmosphere. It takes about 5 times as much energy to evaporate that water as it would take if we heated it all the way from freezing to boiling. That's a lot of energy stored in the moist tropical air. And when that air moves into the cooler regions away from the tropics, the water molecules condense into rain or snow. All that energy is released! The heat energy generated in the tropics from solar radiation is being moved toward the polar regions. We should expect that global warming would make that process more active. And if it happens suddenly, we will get violent weather—the wind will blow.

Hawkins: So with the extra greenhouse effect warming the earth, the earth has to transport more energy to high latitudes and we are likely to have more violent weather.

(Audience) I'm Sean Baxter. Doesn't everybody just get warmer? Or doesn't it just rain more often?

Brandon: It's not that simple. The atmosphere gets more complicated when you add water. If the ground is heated by the sun, the warm moist air is lifted up to where the pressure is lower—where it is cooled until the water molecules begin to condense into cloud droplets. If the wind is blowing the moist air up a mountainside, or if a wedge of cold air is moving under the moist air, clouds will form and when the drops get large enough, we get rain—or snow if it is really cold. Then there will be low-pressure areas with the wind blowing counterclockwise around the storm regions. It's the energy release from that extra moist air that can make the weather violent. And because the earth has to transport more heat from the tropics to high latitudes, the warm moist air from the tropics will meet the cold air from the polar regions more often and storms will be more frequent. And with the oceans getting warmer, there will be more water vapor in the atmosphere so we should expect violent weather more often. The IPCC says that will likely happen. But the weather varies with time and place. Some places might get more storms; other places might get fewer. Remember though, the greenhouse effect is a global thing. It affects the climate everywhere. The daily weather just shows some of the more extreme effects, like floods and droughts and windstorms

Emily: You said that violent storms are likely to happen more often. Does that mean more tornados?

Brandon: Possibly, but that's not an easy thing to predict. For one thing, we are just beginning to learn what happens inside a tornado. It's not very convenient or safe to study them. But we do know the main source of their energy: it's that water vapor that was evaporated from places like the Gulf of Mexico. And the extra greenhouse effect from gases like CO_2 is working to increase that.

Hawkins: With global warming, the average temperature for the entire earth will increase. But the earth will still need to transport heat from the tropics to the poles. And if the ocean transport is slowed by too much fresh meltwater, the atmosphere will have to take up the slack. Violent storms like hurricanes could become much more frequent.

(Audience) Name is Leslie. Brandon said the winds blow counterclockwise about the storms. Do hurricanes always rotate counterclockwise? I think I saw a satellite picture on a book cover of one going the other way.

Hawkins: I think I know the one you mean. But yes, all hurricanes in the northern hemisphere rotate counterclockwise. That's because the earth is rotating; it's called the Coriolis Effect. When a non-scientist tries to illustrate the way we might change the weather or climate, he should be careful about the basic science. The scientific facts are not something that we can change. Maybe it's a good time to remind you that CO_2 is a greenhouse gas because its molecular structure works to make it a good absorber of infrared radiation. That we cannot change. What we should change is the amount of CO_2 that we are putting into the atmosphere by burning fossil fuels.

(Audience) Hi. I'm Corey Manchester. You know, once in a while we get a really big hurricane, like category 5. And some say that's global warming, others say not. Who, or whom?, am I supposed to believe?

Hawkins: Right! There seems to be some argument about the effect of global warming on these storms. Brandon has been studying what we know about these storms. Maybe that will help understand the role of global warming. Brandon—

Brandon: OK. Global warming should produce warmer sea surface temperatures. But the changing ocean currents could also affect that. However, measurements show that all oceans have warmed about 1° F in the last 30 years. So there will be increased evaporation and more energy stored in water vapor—a lot of energy. It's definitely true that hurricanes are the largest powerful storms. They release a very large amount of water vapor energy and transport it away from the tropics as they move to higher latitudes. That happens out in the Atlantic Ocean, but also in the Pacific where they are called typhoons, and in the Indian Ocean where they are called cyclones. The extra greenhouse effect is making the oceans warmer and that will definitely add water vapor energy to the atmosphere. And a deeper layer of warm ocean water just keeps adding energy to make stronger hurricanes.

But of course it's not that simple. If the clouds aren't allowed to grow, say because winds up high blow the tops away, the hurricane will not even get started. Or, if the source of energy is removed as a storm moves over colder water or over land, the storm will not grow. The winds will decrease. All the conditions have to be just right for a hurricane to develop and grow into a big storm. But now we have that extra greenhouse effect working in favor of the big storms.

(Audience) Corey: Can't government scientists do something to stop these storms from happening with global warming?

DAY TWO

Brandon: No. Once we let global warming put that extra water vapor energy in the atmosphere, there's nothing that humans can do. We will just have to move away from the storm tracks, or build stronger houses on higher ground, or maybe move underground!

III. GREENHOUSE GAS INCREASES

Hawkins: The IPCC has said that it's the increase in carbon dioxide and other greenhouse gases that is responsible for global warming. So what is it that controls the amount of CO_2? Back to Cody on this--

Cody: It's obvious that burning fossil fuels like coal and oil is putting carbon dioxide (CO_2) into the atmosphere, and that human activity is increasing those amounts. And we mentioned earlier that plants use the carbon dioxide to grow and put oxygen (O_2) into the atmosphere. We need to think about that some more; the life of a carbon atom is very complicated.

The part that the trees in our forests play is very important. Let's say that we burn some coal and put 100 carbon dioxide molecules into the atmosphere. Some of the molecules are quickly dissolved and stored in the oceans, the trees start using some of it to make wood, and all the other plants are using sunlight and CO2 to grow—so after about four years we might expect only about half of those molecules would be left in the atmosphere. And after another four years, half of those would be used up, but we still have 25 of the original molecules in the atmosphere. We could say that the lifetime of this process is about four years.

Robert: So, sounds like Nature could keep up with the increasing CO_2 in the atmosphere if we just plant more trees.

Cody: But we have to look at the rest of Nature's story. The carbon in those trees doesn't just stay there forever. Maybe 75 years later, the tree dies and falls down. It immediately begins to decay and in this process, many carbon dioxide molecules start to go back into the air; some carbon remains stored in the roots, and a few carbon atoms are used to make rocks. Then in maybe another 75 years the fallen tree has completely decayed. But the molecules that remain in the ground are part of a very large reservoir of carbon compounds that has accumulated over thousands of years. There is a similar reservoir in the oceans. The molecules in these reservoirs have been cycled in natural processes in and out of the atmosphere in large amounts with lifetimes ranging from 1 to 10 to 1000+ years.

Robert: So the forests don't work very fast to control the CO_2. And Nature is very slow in balancing the CO_2 with some of those other reservoirs.

Cody: Right! So this means that every 4 years or so we increase the amount of carbon dioxide in the atmosphere. Some of this carbon is cycled in and out of the forests and other plants back into the atmosphere as CO_2 every 150 years or so. But by burning the carbon stored in those very old reservoirs of coal, oil, and natural gas we are gradually increasing the number of molecules stored in the large land and ocean reservoirs, but the rate at which the CO_2 in the air is going into those large land and ocean reservoirs is very slow. Some of it stays in the atmosphere. Now, say we stop burning fossil fuels. Sure, the extra CO_2 in the air is going into the tree roots and leaves in the soil and finally into rocks, but some of it escapes back into the atmosphere. There is a similar cycle in the ocean. The lifetimes of some of these cycles range from decades to thousands of years. That's why we say that the carbon dioxide increase stays in the environment for something like a century—and it increases the greenhouse effect for all of that time.

Hawkins: So we can make fast increases in the atmospheric CO_2, but once it goes into the trees, the changes slow down. We still have some control of the forests though. But when the carbon goes into those large land and ocean reservoirs, it's out of our control. We can't assign a specific lifetime for that cycle, but it's long—like a century or more.

Cody: That's true. Nature's control is slow. Unfortunately, humanity is putting CO_2 into the air a lot faster than we did years ago. Still, the amount of nature's control depends somewhat on the amount of forest. If we cut the trees for lumber, or to plant other crops to graze cows—and don't replant the trees—the carbon dioxide won't be removed from the atmosphere quite so fast. But, maybe if we increase the amount of forest, we could improve our control.

Hawkins: Do we have a question or a comment from the audience?

(Audience) I'm Corey again. But if the CO_2 increases, won't the trees and plants grow faster?

Cody: True, and some people argue that more CO_2 from burning fossil fuels would be a good thing. There would be more food from plants for all the hungry people in the world. Maybe. But wanting it to happen won't make it happen. The plants would need more water and maybe fertilizer to use the extra CO_2. And extra fertilizer can add more of another greenhouse gas—nitrous oxide (N_2O)—to the atmosphere. And more CO_2 means more global warming. That is likely to cause a lot of problems. It could disrupt the present water situation: you can't grow crops in the desert!

IV. PREDICTIONS OF CLIMATE CHANGE

Hawkins: We've been looking at the observed temperatures and other evidence of global warming, and we have a theory that the increased greenhouse effect of gases like carbon dioxide is the cause. How do we test that theory?

Kyle: Well, we could just keep increasing the CO_2 and see what happens. But if the answer is yes, we would be way beyond the tipping point for disaster. Fortunately, we can use big computers to get the answer; it's a really big calculation problem. All those facts about the solar radiation, the reflectivity of the earth, the amounts and absorption by greenhouse gases, and the feedback effects have to go into the calculation. The feedback effects are really important. For example, if you warm the air, it will hold more water vapor and increase the greenhouse effect. And some clouds could make things worse, or maybe other clouds could make it better. And melting the ice means less reflection of sunlight with a positive feedback and temperature increase. And a really nasty feedback would happen if the temperature increases enough for carbon dioxide and methane to come out of the ocean, or out of the vegetation under the permafrost. That would be a tipping point.

So we have to use fast computers with lots of memory. Even then, it takes a long time to get the answer. And to check the answer, the calculation has to be done for times in the recent past where we have all those experimental observations. Then, if the answer agrees with the temperature facts, the programs can run to make future predictions. And, of course, the predictions depend on assumptions for the future—population, technology, energy sources, that sort of thing. It's different for every spot on the earth too, so the predictions will be different for each continent and for land or ocean.

(Audience) Whitney: Can I see the answer for right here in Arvada?

Kyle: Sure, but the results for the average global temperature will probably be somewhat more reliable. And it's a global problem; your life will depend very much on what happens in other places.

(Audience) Whitney: OK. So what is the answer for the average temperature increase?

Kyle: Well, first, scientists define Climate Sensitivity as the global average temperature increase resulting from doubling the amount of CO_2. Computers provide an answer of about 3° C.

(Audience) Dylan Bradley here. But what's the answer for when I'm out there earning my first million?

Kyle: Unfortunately the IPCC says that if we keep the atmosphere just the way it is today and don't add any more greenhouse gases, the temperature will increase just a little more because much of today's CO_2 will stay in the atmosphere, and the ocean temperatures are slow to change. If we take into account what the future greenhouse gas emission rates are likely to be, the projected warming is expected to be 0.2° C for each 10 years for the next 20 years. (This does not take into consideration possible changes resulting from international emission agreements—and those are really uncertain!)

For comparison, remember that the observed temperature increase for the hundred years before 2005 was 0.74° C and the rate is expected to increase dramatically in the near future. According to the IPCC report, the projected temperature rise for the 21st century, assuming a variety of energy sources, is in the range 1.7° C to 4.4° C, with a best estimate of 2.8° C. If this is the case, the sea level rise would be somewhere between 21 centimeters to 48 centimeters (roughly 8 to 19 inches). These projections include the increased ice flow from Greenland and Antarctica observed in 1993-2003, but these estimates could be high or low because the science is still uncertain. The low number of 8 inches would already cause difficulties for Bangladesh and Florida. And 19 inches would be a disaster for all those coral atolls and a nasty problem for all coastal areas. The possibility of an unexpected sudden meltwater or ice discharge from Greenland or Antarctica would be a pretty awful surprise.

Doubling the CO_2 from a pre-industrial value of 275 ppm would mean 550 ppm—that's parts per million, and a temperature increase of 3° C. Now we hear of a need to limit the global temperature increase to 2° C above what it was before we started burning so much fossil fuel, and hopefully that is a safe goal that will prevent the Greenland glaciers from melting into the ocean, or the carbon from coming out of the permafrost. To stop at 2° C above the pre-industrial background means no more than about 450 ppm. Let's look at that goal of 450 ppm of CO_2: the IPCC says we were already at 379 ppm in 2005 and rising! It's later than you think!

Hawkins: That pretty well wraps up the science. We've discussed the observations, explanations, and the theoretical predictions. Now it's time to look at how we adapt, and look at some possible technological solutions to the problem.

V. SOCIETAL RESPONSE

A. ADAPTATION TO CLIMATE CHANGE

Hawkins: So we've met the enemy—and he is us. That's from the old Pogo comic strip I believe. And I have to tell you my mule story. Probably some of you are familiar with mules—they were useful animals for moving things around before trucks and farm machines. The Forest Service still uses them. They have the reputation for being very smart but very stubborn. Now this is not a true story so don't get concerned about animal rights. This farmer and his friend were about to hitch up a mule to move a wagon some place else on the farm. But first, the farmer picks up this heavy 2 x 4 piece of wood and clobbers the mule between the eyes with it. His friend is horrified and asks, "Why did you do that?" The farmer replies, "Well, shucks, I've got to get his attention first!"

I know, what does this have to do with climate change? Well, if we've got your attention to the problem, now it's time to fix it! We've discussed the evidence and the scientific explanation for the cause of global warming, so now we'll give some more thought to our response. With so much uncertainty about the future, maybe we should just wait and see what happens. Is that an option?

Emily: OK, Dr. Hawkins. Let's consider that alternative. Let's pretend we can ignore what's causing the problem and just adapt to what happens. Sounds easy and nobody expects anything dreadful to happen right away. Don't worry—be happy! We can go with BAU—that's business as usual. Climate change might even be good for some of us. Or who knows, maybe there will be a surprise with something good for the climate. Like maybe the sun will suddenly put out a little less solar energy. Or maybe a big volcano will go off and the dust will reflect more of the solar energy.

(Audience) Jeremy: That doesn't sound like something we should plan on.

DAY TWO

Justin: But climate change will not affect everybody equally. Warmer temperatures and more CO_2 for the plants might even be good for some people. And the sea level rise won't affect us here in the mountains.

Emily: But it will be terrible for Pacific Islanders and people living in low coastal areas like Florida or Bangladesh. The disappearing glaciers won't bother New York City, but there already is less drinking water in Peru and Ecuador and Bolivia, or countries near the Himalayas. Those people will have to move someplace where the changes aren't so disruptive, maybe down the street from you. And there will be places like our Southwest where we can't grow crops anymore. Shopping for food could get to be a problem!

If we wait until things start getting bad, it won't work to suddenly change our ways. The CO_2 is still in the atmosphere and may be past the tipping point where really bad things can happen. It seems like it would be a lot safer to fix it now.

Hawkins: Let's compare this situation with driving a big SUV. You're on a familiar Interstate cruising along at about 90 miles per hour. It's business as usual. You've assumed the climate situation is under control; scare tactics of global warming can be ignored. But the highway patrol pulls you over and assesses a whopping fine for exceeding the speed limit. If you protest, he says, "Ignorance of the law is no excuse!" And business as usual is likely to cost you; ignorance of Nature's environmental laws can be very costly!

Cody: I've read that Greenland has over 600,000 cubic miles of ice; if it suddenly melted or moved into the oceans, it would increase sea level by more than 20 feet. Obviously all of south Florida, most of Bangladesh, the coral atolls in the Pacific, and all other low-lying coastal areas would be submerged. That's what 'past the tipping point to out–of-control global warming' means.

(Audience) Dylan: Good. The Northwest Passage from Europe to the Orient will finally be open.

(Audience) Whitney again. But what about the mammals and sea life that have lived in harmony there? Oh yes—and the native people that coexist there? There are some who don't believe global warming, and there are others who expect someone else to fix it, so there is no need for regulations. Or they think it would be good to have more CO_2 to grow more crops. Those people don't want any regulations or laws to fix it. Or they hope that tipping point could be 50 or 100 years from now—long after they're gone.

Emily: Yes, probably. But the laws of nature are already in force. Then your children could have a nice disaster on their hands—and they wouldn't think very kindly about their parents.

Kyle: And like the ozone holes, there is a potential for nasty surprises. The 2007 IPCC predictions from computer models do not include the developing feedbacks of CO_2 from the permafrost or ocean— or movement of the glaciers of Greenland and Antarctica. The science is not yet precise. Modelers are reasonably confident, however, of predictions for about a decade. That is our grace period for making the necessary payment for climate control. It's just like paying the mortgage or phone bill.

(Audience) Whitney: Well, we know what our mistake is; let's just fix it. We did it with ozone—stopped making the bad chemicals. We just need some international agreements so everybody cooperates.

(Audience) Sean here again. But China and India are putting more CO_2 in the air all the time. Our neighbor next door says we shouldn't have to fix the problem when they are making it worse. Why should we be the ones who stop using cheap electricity from coal-fired power plants?

Justin: But they can use exactly the same excuse. But if we are really smarter and get busy with the research, we might solve the alternative energy problem. And I've heard that some of our top universities are doing just that. Then China could follow our lead—they always have.

(Audience) Whitney here. I was watching that BBC video called "Hot Planet" the other night. They said there might be some new technology to correct the greenhouse problem, like satellite mirrors, or putting aerosols in the stratosphere, or even artificial trees. Think those would work?

Hawkins: That was generally a good video, but maybe that was a little misleading. People are thinking about such solutions, but we really don't yet know how successful they would be. We shouldn't rely on future technological solutions and continue business as usual. Those ideas may become necessary as a last ditch effort—that is, if they work. But the video did emphasize the immediate need to reduce the emissions of greenhouse gases in the first place with conservation and alternative energy sources. Let's talk about that.

(Audience) Keisha: My Mom was having an argument with my Uncle Joe—he's a lawyer. He says there are all kinds of problems of side effects with new technology, even for alternative energies. And some of the new laws that are supposed to solve global warming would probably cost more and might not work, so we should go real slow. Mom says just pay attention to whether we emit less CO_2. She says those other mistakes can be fixed later.

Justin: Yes, there will be mistakes that will have to be fixed, but the ultimate mistake is to do nothing.

(Audience) Jeremy Sorensen again. If we made all the changes we needed tomorrow, we won't get cooling—things would stay about the same. And 20 years from now might be too late. Climate change might be past the tipping point with no way to come back. So it's hopeless.

Emily: Like we've said. Do nothing and see eventual disaster. Procrastinate and it will cost even more money to fix it or adapt. But the best science predicts the expected increase of CO_2 to 450 ppm—that's parts per million—will increase the average temperature by $2°$ C above the level it was before industrial activity. If the Chinese follow our example of making CO_2 , maybe the average temperature could go

3° C above that level. Then, if you are living near the ocean, or in a place where it stops raining, things could get uncomfortable and expensive to fix.

(Audience) Alberto Jensen here. So just move to Denver.

Brandon: But some of our scientists are worried about unpleasant surprises that aren't well understood and aren't in the IPCC predictions--like movement of Greenland glaciers, and methane and carbon dioxide from permafrost or oceans. Those things are likely to happen if it gets warm enough— but we don't know that tipping point. Denver could suddenly get very crowded!

DAY THREE

{THE OZONE DISTRACTION}

Hawkins: Today we will continue to discuss society's response to disruptive climate change with solutions. Our concluding panel has Jessica Stone, Joshua Jackson, Donald Forsyth, Benjamin Martinez, and Madison Fredericks.

(Audience) I'm Jacob Mason. My Dad says not to pay attention to this talk of global warming. He says there was lots of fearful talk about the ozone, and they fixed that. That happened before we were born. But we still get ozone alerts so maybe there is still a problem.

Jessica sat with her mouth open. Jacob, and perhaps others in the audience, had strayed away from the climate change problem. Global warming was being confused with the stratospheric ozone problem of past years. There was obviously a misunderstanding. Kyle glanced questioningly at Dr. Hawkins.

Hawkins grimaced and grinned slightly. "Yes, that's a different problem; you'd better explain that if you can. I'll help if you need it."

Jessica: Well, ozone was a different atmospheric problem. The damage to the stratospheric ozone layer was a serious global problem because if it didn't get fixed, people everywhere would be exposed to more damaging ultraviolet. Anyway, we did correct our mistake with that, didn't we Dr. Hawkins?

Hawkins: Yes, the damage to stratospheric ozone was caused by molecules containing chlorine, called CFCs, that were used for air conditioning and other things. We've stopped putting CFCs in the atmosphere. And the ozone alerts that you mentioned are a separate problem that has to do with pollution in the lower atmosphere, especially from automobiles. Then energy from bright sunshine starts some chemical reactions to make ozone. It's bad to breathe ozone; it burns the throat and lungs. And we've been working to fix that problem. But both of these ozone problems are different from global warming. Go ahead Jessica.

Jessica: OK. Global warming comes from the extra greenhouse effect caused by molecules like CO_2 that our industries and automobiles are putting into the atmosphere. This happens when we burn fossil fuels to make energy. It's the increased greenhouse effect in the infrared that is making the earth get warmer. It's a different process from the ozone absorption of the ultraviolet sunlight. It's the same atmosphere, of course, but it's a different problem. We're just learning about it, and we still need to fix it.

Hawkins: We've mentioned that other atmospheric problem of ozone. It is different from global warming too, but maybe we can learn something from it. What do you think, Joshua?

Joshua: Yes, I have a history lesson for you. We won't remember it because it happened before we were born. Ozone is a molecule made with three atoms of oxygen--ordinary oxygen that we breathe has only two atoms. Anyway, the ozone layer in the stratosphere absorbs must of the ultraviolet that can cause sunburn and sometimes skin cancer. But our parents' generation manufactured and used a compound containing chlorine for air conditioning and other

stuff. Some scientists became worried that it would damage the ozone layer in the stratosphere. Damage to the ozone layer would affect everybody; it was a global problem. And suddenly it became a big problem that couldn't be ignored because scientists observed a big hole in the ozone at the South Pole in the springtime. That was a surprise. But then, because everybody realized it was an important global problem, there were immediate international agreements to correct our mistake. All the countries stopped making and using the stuff containing chlorine molecules. But we can't just go back to square one overnight, because those old molecules with chlorine are still floating around out there. We can't just scoop them up out of the atmosphere. They will die away slowly, and the springtime ozone holes will finally disappear in maybe 50 or 75 years.

Donald: So, what does this have to do with global warming?

Joshua: It's just a lesson that we have to learn: stop making the same kind of mistakes. First, we need to be aware that there may be some unpleasant surprises out there. And we shouldn't procrastinate about fixing our mistakes because we might have to live with them for a very long time. And since climate change is a global problem, all countries need to cooperate on fixing it, just like they did for the ozone problem.

B. *SOLUTIONS TO CLIMATE CHANGE*

Hawkins: Back to today's problem of Climate Change. We've talked about adaptation. What are the solutions? The IPCC has said that it's the increase in greenhouse gases produced by humans that is responsible for global warming. We fixed the ozone problem by stopping the production of the CFCs; we found substitutes to use in air conditioning and spray cans. But decreasing the greenhouse gases is not so simple. For example, some of the atmospheric methane (CH_4) comes from cattle and some nitrous oxide (N_2O) comes from the fertilizer we use to grow plants. We can't just stop producing food though; we will need even more food for increasing populations. But we might change or improve our agriculture practices. The increased CO_2 is an even bigger problem; most of the increased CO_2 comes from burning fossil fuels for energy—but our civilization needs lots of energy. We see that trees work to remove the CO_2, but not fast enough. The obvious way to control the CO_2 would be to use less energy or to stop burning fossil fuels to produce it. Fixing this global warming problem isn't going to be as simple as fixing the ozone and ultraviolet problem was. We can't just suddenly stop burning coal with its emission of CO_2 and switch on nuclear reactors and wind turbines.

So what are some possible solutions? Let's have an open panel discussion about solutions or how we adapt. Let's open this discussion to the audience—a real town meeting. Any questions or comments would be good. Our panel will try to give replies.

DAY THREE

(Audience) Hi. My name is Alberto Jensen. Why aren't we using alternative energy? If it's so easy, why haven't we done it? I know—we've got all these other expensive things to do, like a war or two and medical bills and taxes for schools. People say we can't afford it just now. So what happened to all those dollars we saved by using cheap energy and causing this climate problem?

Jessica: True, our parents have been a little slow to see the problem. And there are lots of other things to worry about, like terrorism and how everything is getting more expensive and a lot of people can't find jobs. So some people may think that maybe this isn't so urgent and it can wait to be fixed sometime in the future, a generation or two from now. But the science is a definite. Maybe it's time to see that this is as if we've been feeding a big monster that has arrived on the earth and is about to take over; all the other things won't seem so important. It's time for everybody to work together on this!

(Audience) Keisha Martin here. I hear about all these environmental problems. Like first it was DDT and the eagles, then acid rain and the fish, nuclear testing and fallout, CFCs and ozone, now CO_2 and global warming—all because of technology. Can't we ever do good things in science?

Jessica: Those were some of our worst mistakes, but we used technology to correct them. Take coal-burning power plants—one of the worst sources of CO_2. We don't have to stop burning coal to make electricity. If we stop saying 'ain't it awful' and work on the technology to capture the CO_2 before it goes into the air, then store it in the ground or the ocean, we will be doing a good thing. Might take some real work and thought by our scientists, and electricity might cost more, but we could score big doing that—a better way to spend our money than just building another old-fashioned power plant that spews out CO_2.

Robert: Well, we already know how to do nuclear power. Now we just need to let technology do a good job on all the safety stuff and play catch up. And nuclear fusion would be even better—there's no radioactive waste. Fifty years of research on fusion and we are getting closer, but slowly. We have just been lazy and used cheap oil and coal and emitted a lot of CO_2.

(Audience) Michael here again. Nobody wants the nuclear waste in their backyard. And with all the red tape and building time, the ocean will flood New York City before they throw the switch!

Benjamin: That's a good point. So we have to do some other things too. For example, the sun sends thousands of times the power we can generate from coal. We use just a small part. We could heat water with solar panels on the roof and we could generate electricity with solar cells like the ones on the space station. The sun makes the wind blow, so we could put up wind turbines to make electricity.

(Audience) Dylan: Let's say that I know there is a problem and I work to fix it. I change light bulbs and try to do my part to conserve energy. But the next step isn't easy or cheap—like solar panels. If

DAY THREE

a CEO says switch from electricity from coal to a wind farm, it costs a little more of course. His stockholders—those who earn money from the coal company—will complain. If it were easy, why aren't more folks doing it to get to the $2°$ C goal?

Madison: A few are. They find a way to do things more efficiently or expect a future profit. But not fast enough! Some say we're dangerously close to the tipping point, whatever that is.

(Audience) Jeremy: Yes, stop putting CO_2 in the air. So we stop burning coal and the lights go out. Or we stop driving our cars—forget soccer. And grow all our food in the back yard. You're talking about miracles here!

Hawkins: Sounds difficult, doesn't it. And like the CFCs, the CO_2 from yesterday won't go away anytime soon. And like CFCs, we can eventually fix our mistake, but getting international agreement on solutions isn't prompt like it was for ozone. The solutions won't be easy.

Joshua: Right. The time to correct is now. So do it! We need to get our energy for lights and travel from something besides fossil fuels.

(Audience) Right! I'm Tyler Swenson. Seems to me that once we stop being distracted by little differences like keeping up with the Joneses, or who has the biggest army, we can lead the world in solving this common problem with the laws of Mother Nature. And if we don't sit on our hands too long, we might be able to do wind and solar faster than nuclear, and maybe fix some of those coal plants to slow down the CO_2. And the poor nations in Africa and the

Pacific who already have problems will need more help in adapting. We have to do all those things.

(Audience) Sean here. But won't it cost a lot?

Joshua: Probably at first, just like it did for sending a man to the moon. And look at the computer industry. It could happen something like that. But the more we delay, the more it will cost. We need to look ahead to a future where we've finally gotten smart and worked hard to make things right!

Hawkins: Do you have a suggestion, Madison?

Madison: Well, we waste a lot of energy. Almost a third of the energy we use today is used in buildings. They could be better insulated. And all buildings could use these special fluorescent lights instead of filament type—that would save a lot of energy. Right now, the bulbs are more expensive, but they last longer.

Joshua: Actually, Congress has passed a law to stop making the old ones. Everybody will have to buy fluorescents. No financial loss for industry because the same companies will manufacture the new ones.

Jessica: Even more energy goes into travel. A lot of people drive big SUVs at 75 mph; they burn lots of gas that puts CO_2 in the air. But I know people think they are safer.

Joshua: And that same law says that the automakers have to make cars that are more efficient. The cars will probably be smaller with more efficient engines. And it's speed that makes any car less efficient and more dangerous—people should just drive more slowly. We won't need those heavy cars to be safe. Everyone should try to drive less—maybe half as many miles as they do now.

Madison: New York City is trying to have taxis that are hybrid—electric with small engines to charge a battery. And London charges for driving big cars into the city. Denver could have more public transportation like busses and tri-rails, with lots of small electric taxis at the stations to take people right to their office door. There are actually some places like Israel that are doing just that. That should improve the efficiency in commuting by stopping the CO_2 and other emissions from those big SUVs. It doesn't eliminate the CO_2 emissions of course—just moves it to the electric power plant. But if we can figure out how to make carbon free or carbon neutral electric power generators we would be in good shape. No reason to have SUVs and six-lane highways for commutes.

Jessica: There's also a lot of air travel. And those jets have to burn liquid fuel that emits CO_2. But too much of air travel is just for business. Companies could have web conferencing or video conferences instead of traveling.

Benjamin: And there are hidden ways that we increase CO_2. It takes money and energy to transport the food we eat—or to send food to people in poor countries. And we need to look at the energy to transport raw materials to build new power plants or the fuel to run them. We need to pay attention to that part of the problem, too.

Hawkins: Efficiency would be good, but will that be enough?

Robert: I agree, everybody could be more efficient— shutting off the lights, driving smaller cars, that sort of thing. But we need to produce energy some way that doesn't increase the CO_2.

Hawkins: As you probably know, at the present time the United States and China generate electric power mostly from burning coal because there's lots of it—and it's the cheapest way to go. So it's a major source of the increased CO_2. What do we do about that?

(Audience) Alberto: Well, some folks argue that we just have to continue on that path for economic reasons and hope the global warming problem will somehow go away. And when we need more electricity, the power companies just plan to build another old-fashioned coal-burning plant.

Jessica: That's just procrastinating and trying to ignore the problem. We've learned how to remove some of the pollutants, like sulfur compounds, from the smokestacks, but it's still not clean coal.

It should be possible to capture the carbon dioxide as well. But there's been little effort in that direction because it will cost more.

(Audience) Dylan: It seems to me that global warming and disruptive climate change are clear and present dangers. I think we should use our limited time and money to develop alternative energy. Or the people who think we need to continue to use abundant coal should get busy and work to capture and bury CO_2.

Madison: Well, solar energy is carbon free, and it's renewable. I've read that the amount of solar energy possible is thousands of times greater than all the electric power generated on earth.

(Audience) Whitney again. So why don't we use it?

Madison: That's a real good question. We know how to do it. The trouble has been that with cheap fuel like coal and oil, it was more expensive. So we don't have lots of inexpensive wind turbines and solar panels all ready to go. It will cost more to get started.

Jessica: But since the solar radiation and wind is free, everybody could invest a little and in just a few years, we would come out even. And we would be solving the global warming problem.

Madison: It's true that most people don't have a lot of ready cash to spend for solar panels and American industries haven't started mass production to make them cheaper. Business people are afraid to invest in factories if there's no guarantee of a future market. Some countries like Germany and Japan are way ahead of the United States in making this transition because there are government financial incentives to make it happen immediately and a promise to make it the policy for the future.

Hawkins: Isn't nuclear fission carbon free? Robert?

Robert: Yes, nuclear energy certainly is. But there were a couple of accidents and got everybody scared of it. And the waste products are radioactive. So NIMBY—not in my back yard.

Jessica: France generates most of their electricity using nuclear power—and they have never had any accidents. I guess if you do the safety things correctly, it shouldn't be a problem. If people are really worried about global warming, this should be part of the solution. We don't see it happening though.

Robert: That's because it takes a while to get started—design, location, permits—all that stuff.

Benjamin: Nuclear fission has that problem of radioactive wastes. But what about controlled nuclear fusion? Its by-products aren't radioactive. Research started on that 50 years ago, but it did not develop very fast because we had cheap coal and oil. Now there's a new international facility being built in the south of France. If that works, controlled fusion might be a good long-term solution—but not soon enough! And I hear that Congress isn't sending our share of the money to help build it.

(Audience) Alberto: I've heard that Hansen from NASA says we have a grace period—like for paying telephone bills—of only about 10 years before we would go past the tipping point and have bad surprises. Reactors won't be ready that fast.

Hawkins: So what about burning wood and other plant materials—call it biomass—to make energy. Joshua?

Joshua: Yes. That would be a renewable energy source. And replanting the trees and plants would make it carbon neutral. By that, we mean that the plants would remove the CO_2 as fast as it is being emitted. Wood can be burned to make energy of course; we heat our mountain cabin that way. And there's even a carbon negative possibility. We could keep the trees growing, then burn the wood to get energy, but capture the CO_2 before it goes into the air. Or just bury the trees after they have taken CO_2 out of the air. That sounds like a stupid solution though. And crops like sugar cane and corn and palm oil can be made into something called ethanol-- that is a liquid fuel.

(Audience) Michael: Right now ethanol is being mixed with gasoline to run automobiles so we don't have to depend on so much imported oil.

Joshua: Yes, lots of talk about not having to import so much oil. But that's a different problem. For the really important problem of global warming, the whole point is to be carbon neutral by using a renewable energy source.

Benjamin: It takes energy to transport that oil too. So it's a lot better to grow a renewable fuel close to home.

Joshua: OK. But, the soil and water have to be right to grow the renewables. We can't do it with poor soil or lack of water, like the deserts in the southwest. And the sugar cane and corn can't be used for both energy and food. Food could become scarce and more expensive. Producing ethanol to run cars could mean increased numbers of hungry people in the poor countries. Corn meal for tortillas will cost more and palm oil to cook with will be hard to get. We need to look at all those side effects too. Biomass can be some help in controlling global warming, but it can't be the only solution.

Robert: Why don't we burn hydrogen to run our cars and trucks? Then the exhaust is just water vapor— no CO_2 at all! They're trying it in Iceland, and since their electricity is produced geothermally, they can be carbon neutral.

Joshua: Good solution! We would just have to switch from gasoline stations to hydrogen stations and get somebody to furnish the hydrogen. It could be done. We can make hydrogen by electrolysis—we just have to solve how to generate electricity without emitting CO_2, then full speed ahead!

Robert: Or we could burn wood to heat water for steam like the old locomotives—or cars and trucks, like the Stanley Steamer! And replant the trees, of course.

Joshua: I'm not sure civilization wants to go that far back!

Robert: Right. So the time to correct is now. Do it! We need to get our energy for lights and travel from something besides fossil fuels.

Hawkins: I wonder if the failure to act on global warming isn't a bit like the eating habits of some of you. You're in the habit of eating junk food and guzzling soft drinks. This is somehow related to your extra weight and the flab around your waist. And you've read someplace that your arteries are getting clogged and that eventually you'll have some nasty medical problem—or get a heart attack. But there are all these cute TV ads and billboard pictures that say not to worry. And junk food is so cheap and easy!

OK, I'm ready to hear some final comments on what to do about disruptive climate change. We can keep adapting to the problems that are likely to happen tomorrow, or maybe we should fix things for the future—pay now for free clean energy from the sun and wind a few years from now. I see a bunch of hands up in the audience. Let's start right there in the front row.

(Audience): Whitney again. Seems to me the main problem is: We've got to have energy to keep the machine going. You know— civilization. But we've got to get control of those greenhouse gases before we have a wreck.

(Audience) I'm Samuel Barnes. So we need to make energy without burning fossil fuel; and we can do it!

(Audience) Sean again. And the geeks with the smarts to invent the technology can get us organized to do it right. There are lots of ways to be more efficient with transportation; no need to sit in traffic making more greenhouse gases—and all kinds of other pollution.

(Audience) Cody here. Remember me? If we get busy building solar panels and windmills to make clean calories, there will be a bunch of new factories and jobs.

Hawkins: There's a very excited young man in a Bronco sweatshirt in the back row jumping up and down and waving his arms. Let's see what he has to say.

(Audience): Name is Leroy. I'm not at the top of the class; Mrs. Murphy says I'm smart, but lazy except on the basketball court. But I hear you loud and clear. So—we got us a problem. And the longer we wait, the worse it gets. It's the CO_2 stupid! Well, let's fix it!

Joshua: Seems to me that Leroy has it right, all the way down to the bottom line! We all need to recognize that global warming is the world's common enemy. Too many folks are distracted by the economy and international affairs; climate is the biggy! We have

not been paying for the climate damage from burning fossil fuels; now the problem is coming home to roost! We'd best use that alternative energy that's just sitting there waiting.

Jessica: Right! Just sitting on our hands isn't helping. The problem just keeps getting worse and adapting to it gets more and more costly. Our wonderful technology has caused the problem; now we should use it to repair the damage. Sure, it's a global problem, but we should lead the way to preserve the life we've been given. If we don't accomplish anything else in our future, we should work to save the planet.

Hawkins: I find this response from your generation very impressive. And with your future leadership, the world will most certainly follow!

I would like to add in conclusion that we are tempted to think that the earth's climate was designed just for our benefit. But today's science indicates that we are causing it to change. We may hope that some supreme being, perhaps Gaia, will continue to adjust the climate for our benefit, but I believe that humans, as an intelligent favored species, should accept the obligation to be good stewards for the planet. And sound logic says that we should work to maintain these climate conditions favorable to our existence!

SECTION B. THE SCIENCE OF GLOBAL WARMING AND CLIMATE CHANGE

I. Global warming—the temperature record

Global warming means that the average temperature of the earth is increasing. Temperature is a measure of the random energy of motion of molecules in the air. Molecules of nitrogen and oxygen are moving with a range of speeds as they bounce about, colliding with each other. A high number for the temperature means the molecules are bouncing around at high speed. We say it is hot. A low number means it is cold. We wear appropriate clothing and adjust the heating or air conditioning for our personal comfort. All thermometers must agree on the temperature for the same situation. In the United States, we commonly use the Fahrenheit scale. This scale must read 32° when water freezes and 212° when water boils (at sea level pressure). Many other countries use the Celsius scale that must read 0° and 100° for those conditions. There are 180 divisions between freezing and boiling on the Fahrenheit scale, and 100 divisions on the Celsius scale. This means that a change of temperature of 1 degree on the Celsius scale (1° C) is a change of 1.8 degrees (1.8° F) on the Fahrenheit scale.

The Intergovernmental Panel on Climate Change (IPCC), a large group of scientists from many countries, displayed the yearly record of global average surface temperature from 1850 to 2005 in a report in February 2007. An average surface temperature is calculated from measurements at each location on the earth for each season. The surface temperature reported for 2005 was about 14.5° C. This is 14.5° C above freezing on the Celsius scale, or 26.5° F above the freezing point of 32° F on the Fahrenheit scale. The Fahrenheit temperature would then be $26.5 + 32 = 58.5°F$. From this information, the scientists conclude that the average temperature of the earth has increased by 0.74° C in the hundred years ending in 2005. This is a 1.33° F temperature increase. A temperature record of a slight trend with considerable variations would appear to be a rather insensitive indicator of climate change, but the presence or absence of ice and snow, or water, change significantly with temperature. The evaporation of ocean water to atmospheric water vapor is also influenced by temperature, and atmospheric water vapor plays a significant role in climate change. Animals and plants have adapted to the average temperature and range of conditions for each place they live and in some situations the changes may very likely have disruptive effects.

II. THE GREENHOUSE EFFECT

We hear that the "greenhouse effect" controls the earth's temperature. So how does a greenhouse work? These buildings are frequently used to grow vegetables or flowers under controlled temperature conditions. They are constructed with glass or plastic roofs that allow sunlight to enter for heat and light for the plants. When the glass or plastic blocks the heat radiation from the plants and ground, the heat is trapped inside. Actual greenhouse temperatures are usually modified by supplementary heat from local sources or by discharging excess heat to the outside with fans or open windows. Planet Earth does not have the option of removing excess heat; the temperature is controlled entirely by the balance of visible and infrared radiation.

SECTION B

To understand the earth's greenhouse effect, we need to know some details of solar radiation. The radiation is strongest—the sun is brightest—in the visible part of the spectrum, where our eyes are sensitive. We see colors from violet, to blue, to green, to yellow, to orange, to red in the order of increasing wavelength. But there is ultraviolet at shorter wavelength beyond the blue and violet. The ultraviolet intensity is less than the visible, but it is concentrated in packets of energy that are capable of producing sunburn. The longer wavelength region beyond the red, called the infrared, is important for the greenhouse effect. The intensity is less, but the spectrum extends a great distance to long wavelengths. Although our eyes are not sensitive to infrared radiation, we can detect the heat when it is absorbed. If the infrared is intense, we can feel it on our skin. There are scientific instruments of various sorts that can detect and measure any of the radiation.

The solar spectrum is a continuous distribution of energy with wavelength, except for narrow sections that are absorbed by molecules on the sun or in the earth's atmosphere. The physics theory of this spectrum is well known. The radiation from charged particles in the light source is distributed according to their energy of random motion, that is, the temperature. (Here it is convenient for scientists to use a temperature scale called the Absolute, or Kelvin scale. It uses the divisions of the Celsius scale, but shifts the zero point to -273° C where the random motion of the molecules stops.) The heat spectrum is determined entirely by the temperature, independent of the material of the light source. The temperature of the sun's surface is about 5000° C; the maximum intensity is in the visible. But according to the theory, and as observed, the maximum intensity of radiation from an object near room temperature, say 20° C (68°F), is shifted to the infrared. And with our hand, we can feel the heat from the absorbed radiation, decreasing from that of a white hot object, to a slightly cooler red hot at 4000° C, to that which looks normal but is only slightly cooler, say 2000° C, or even as cool as a few hundred degrees.

So, the heat radiation determined by the temperature of the warm ground and plants in the greenhouse is the infrared and is absorbed by the glass in the greenhouse. If it gets too warm, we open the windows. On the other hand, the temperature of the earth is controlled entirely by the balance of the radiation; we can't open the windows to let extra heat out to space. NASA has a satellite that is measuring the total incident solar radiation. We can do a calculation, balancing this incoming radiation with the outgoing infrared radiation from the earth. Since this theoretically depends only on the temperature, we can calculate what the average global temperature of the earth should be. If we ignore the reflection of some of the solar radiation and leave out the atmosphere, we find the answer to be 7° C, but it is -7° C when corrected for reflection from the ground. But that is 21.5° C below what is observed. Why? Because we have omitted the greenhouse effect. And we haven't even considered the reflection from clouds.

The earth has no glass window to trap the infrared, but it has an atmosphere. There are molecules like water, carbon dioxide, methane, and others—we call them greenhouse gases—that absorb strongly in the infrared part of the spectrum, much like the glass in a greenhouse. These molecules are responsible for trapping sufficient heat to maintain the earth at a temperature for life as we know it. But if too much radiation is trapped, we will not be able to vent the excess heat out a window to space. Our concern is the increase in these greenhouse gases caused by human activity that produces additional heating by the greenhouse effect. The temperature response to the increasing energy trapped by the greenhouse effect can be complex. While there continue to be uncertainties about many details of the earth's system of natural laws, there can be no doubt about the fundamental science of the greenhouse effect. The probability of disruptive climate change should not be ignored.

III. THE GREENHOUSE GASES

The IPCC says that the greenhouse gases carbon dioxide (CO_2), methane (CH_4), and nitrous oxide (N_2O) have increased markedly as a result of human activities. Their report indicates that CO_2 had increased from a pre-industrial value of about 280 ppm to 379 ppm in 2005. Scientific measurements for 2009 show a continued increase to 387 ppm. The primary source of the CO_2 increase is the burning of fossil fuels coal, oil, and gas. Similar increases in atmospheric concentrations of CH_4 and N_2O are at least partially a result of agricultural practices. These molecules are broken down into other compounds in the atmosphere by photochemical activity. A large fraction of the methane disappears in about 10 years, but nitrous oxide persists in the atmosphere about 10 times longer. It is fortunate that the increased methane has a relatively short lifetime in the atmosphere because it is a particularly strong absorber of earth's heat radiation. However, a final product of its chemical breakdown is CO_2, a more abundant and more persistent greenhouse gas. It should also be noted that methane is the principal ingredient of natural gas; when natural gas is used for fuel, the carbon reappears in CO_2.

Approximately 70% of the greenhouse effect of gases increased by human activities is due to CO_2. Its environmental behavior is very complex and we need to understand its behavior if we are to manage its role in global warming. When we burn fossil fuels like coal, oil, and natural gas to generate electricity, heat our buildings, or fuel transportation, the CO_2 goes directly into the atmosphere. But when we examine the carbon cycle we find that the CO_2 can never be finally eliminated. True, Nature immediately begins to remove it by plant growth (photosynthesis) and dissolving the gas in the ocean surfaces. In fact, the measured concentrations show a small annual cycle due to seasonal plant growth. The time scale for these processes to remove some of the original CO_2 molecules from the air is only about 4 years, but there are additional natural cyclic processes that cause the actual concentration to decrease much more slowly. The current rapid rate of CO_2 emission into the atmosphere by oxidation (burning) of any carbon substance, including fossil fuels, is much greater than the above loss processes. The atmospheric CO_2 concentration is increasing rapidly.

Let us consider the natural cycle of CO_2 in the growth and decay of the forests. CO_2 is removed from the atmosphere through photosynthesis by a tree. That is, the tree uses the sun's energy of radiation to convert the atmospheric CO_2 into carbon compounds in the wood. This process also releases oxygen (O_2) back into the atmosphere. Say a tree removes the CO_2 as it grows for 75 years before dying; then it decays for another 75 years, putting much of the CO_2 back into the atmosphere. The timescale of the growth and decay of earth's forests imposes a significant lag in the buildup and decay of atmospheric CO_2 produced by the burning of fossil fuels. The concentration in the air will be modified by amounts and rates that will depend on the size of our forests. Obviously, the trapping of CO_2 in the forests is decreased by deforestation and wildfires, significantly worsening the problems of global warming.

Some of the carbon from the CO_2 dissolved in the ocean is stored in the plankton debris in the deep ocean and in coral reefs. On land, trees have temporarily trapped some of the carbon in roots and leaves that have mixed with the soil, and some of it is eventually made into limestone. Carbon stored in these reservoirs will increase very slowly with increased atmospheric CO_2; the time scale ranges from decades to thousands of years. These very large carbon reservoirs also have the potential for recycling large amounts of CO_2 back into the atmosphere as the global temperature increases.

There are many natural effects on the rates of these carbon cycles that are not under human control. With the rise of atmospheric CO_2, ocean acidity is increasing, slowing coral growth and other biological storage of carbon. At some level of future global warming, there may be uncontrolled feedback effects that could accelerate CO_2 emission from the ocean or land reservoirs, abruptly increasing the greenhouse effect and the planet's radiation imbalance. These include release of the CO_2 dissolved in the oceans and melting of the Arctic permafrost with exposure of organic material that produces methane and carbon dioxide greenhouse gases. The equilibrium of carbon storage in our forests may be upset by decreased precipitation and increasing insect destruction or wildfires as temperatures increase. Slash and burn practices in agriculture, as well as wildfires, lead to increasing black carbon deposits on the snow at high latitudes, resulting in increased absorption of incident sunlight and resultant snowmelt. An additional feedback occurs in the reduced reflection of sunlight in the radiation balance. And of course, warming of the oceans increases evaporation and the amount of water vapor greenhouse gas. We are just beginning to understand some of these natural processes, and there may be additional natural surprises that are out of our control.

Finally we might imagine that carbon from plants could follow that very old pathway into reservoirs of coal, oil, and natural gas. But this time scale is so very long that we need not consider it in our climate change for humans.

The current situation may be summarized as follows: We are removing carbon from that very old extremely large fossil reservoir to generate energy. The CO_2 produced from burning this fuel immediately enters the atmospheric reservoir where it functions as a greenhouse gas to shift the radiation balance towards a warmer planet; the concentrations quickly reach a temporary balance by plant growth and trapping in the ocean surface. These CO_2 concentrations are at the same time being modified by carbon trapping in the forest reservoirs. This relatively slow rate is determined by forest growth and decay during our lifetime. Finally, at a much slower rate spanning many centuries, the carbon originally stored as coal, oil, and natural gas may eventually be sequestered as calcium carbonate on land or in the deep oceans. These combined natural methods of balancing the level of atmospheric CO_2

are inadequate to cope with our increasing use of fossil fuels and consequently increasing level of CO_2 production. The amount of greenhouse gas and global warming tends to follow the rate of burning of fossil fuels. The present rate of global temperature increase is the result.

So we find that the earth's carbon cycle is rather complicated with a range of time scales. It is the fast time scale that poses the imminent dangers of climate change. The rate at which the CO_2 is sequestered in the forests is subject to the natural rate of tree growth and decay. And the rate that atmospheric CO_2 can be returned to the normal concentrations of the recent past will be limited by the rate of ultimate trapping in the ground or the deep ocean reservoirs. Even if we abruptly cease the emission of CO_2 from burning fossil fuels, the concentration of atmospheric CO_2, and its greenhouse warming, will not return to normal for many centuries. Human societies could quickly modify the rate of atmospheric CO_2 increase from the burning of fossil fuels. The controlling policy will be determined by scientific knowledge and moral responsibility in conflict with immediate demands for energy and the global economic status.

Considerations of the carbon cycle indicate that natural process rates involving the large land and ocean carbon reservoirs are beyond direct human control. There is a real and present danger that we might procrastinate beyond the time when we might act. Nature will then take control.

IV. CLIMATE CHANGE – OTHER EVIDENCE

A temperature increase from 57° to 58° is hardly noticeable. But somewhere else on the planet the temperature will change from 32° to 33°. (We're using the Fahrenheit scale here.) That makes the difference between ice and liquid water. This may make the difference between rain or snow precipitation and melting of Arctic Ocean ice or the ice caps of Greenland or Antarctic. Sea surface temperatures have increased by 1 degree which results in more water molecules in the liquid escaping to become water vapor molecules (more greenhouse effect) in the atmosphere.

The processes of melting ice and evaporation involve large amounts of energy. Melting ice requires almost as much heat as heating water from freezing to boiling. The change of state of earth's water from ice to liquid is a clear indication of the planet's energy increase caused by the imbalance of radiation. The amount of energy required to melt a certain amount of ice is equal to the amount that would heat the same amount of water from 0°C to 80° C, almost to boiling at 100° C. (Middle school students are able to do this experiment.) The greenhouse effect is increasing the earth's energy by a large amount when it melts the ice. And evaporation from the warm oceans stores an immense amount of energy in the water vapor of the atmosphere that may be released suddenly in violent storms as the earth transports heat from the tropics to polar regions.

So, is the ice melting? Yes! Almost all the glaciers in the world are melting or sliding into the ocean faster than they can be maintained by precipitation. The Greenland glaciers from the ice cap are melting and moving rapidly. That's a lot of energy increase happening right before our eyes. This is extremely worrisome because the ice cap contains the equivalent of 20 feet rise of sea level. This is not likely in the foreseeable future, but the potential for continuing sea level rise is very large.

Sea levels have risen 17 centimeters (6.7 inches) in the past century. That rate was about the same until 2003 when the rate nearly doubled to 3.1 millimeters (0.12 inches) per year. About half of this is due to expansion from increased temperature and the rest is meltwater from glaciers and the ice caps of Greenland and Antarctica.

The Arctic sea ice in summer is becoming less thick and smaller in area. At the end of summer in 2007 the area of summer ice was only half that observed 50 years ago. The open water reflects less of the incident sunlight than the ice; this shifts the radiation balance towards increased warming. The animals, especially polar bears and walruses, are threatened, and native people are having difficulty in adapting.

While the global temperature record may seem erratic and rather insensitive to the greenhouse gas increases, it is clear from the above that there have been extraordinary increases in energy on the surface and atmosphere of planet earth. Changes in life on this planet are inevitable.

V. PREDICTIONS OF GLOBAL WARMING

We described a simple calculation of the balance of incoming radiation from the sun and the outgoing radiation from the earth to demonstrate the importance of the greenhouse effect. A more elaborate calculation requires a detailed account of the infrared absorption by the greenhouse gases. This means that the amount of absorption by a molecule at every wavelength and the atmospheric concentrations of these molecules must be known. This in turn means that details of the incoming solar spectrum and reflectivity of the earth's surface and atmosphere must be utilized. All of these values vary with location on the earth and the calculation is long and complex. Only the fastest computers with very large memories can be used: even then, the calculations take months. The result can only be considered correct if it agrees with ground truth—the observed past and present temperatures. Then, and only then, predictions for future climate change can be made, with various assumptions of future changes in greenhouse gases. Additional complexities arise from what are called feedback effects. A warmer ocean means more water vapor—an increased greenhouse effect—or more clouds. Melting ice means less reflectivity of incoming radiation—a positive feedback of increased warming. The results must be continually compared with updated ground truth observations. Answers different from the measurements must be abandoned. Agreements with observations indicate correct calculations using current science. A very large research effort is required for both observations and calculations.

The climate sensitivity to the increase of CO_2 greenhouse gas is shown by a calculation of the global warming expected from a doubling of CO_2 concentrations with respect to the background level of 280 ppm in the pre-industrial age of 1750. The result is about 3° C (5.4 °F) for CO_2 concentration of 560 ppm.

SECTION B

The IPCC report predicts a temperature increase of 0.2 °C (0.36 °F) per decade for the next two decades for a likely range of greenhouse gas emissions. A long-range prediction for the year 2100 is in the range 1.7 °C to 4.4 °C (3.06 °F to 7.92 °F), with best estimate of 2.8 °C, assuming society's adoption of a variety of energy sources. That's about 5 °F increase by the end of the century. Even if the greenhouse gas concentrations are held constant at present values—that is, with no additional emissions--some additional warming of 0.1 °C (0.18 °F) per decade is expected because of the slow response of the oceans and the very long lifetime of CO_2 in the environment.

Observations of sea level rise have shown a recent increase from 1.8 mm per year to 3.1 mm per year. Sea level rise for the 21st century is predicted to be in the range 21 centimeters to 48 centimeters (roughly 8 to 19 inches). This prediction includes the earlier observed rates of increase, but allows for substantially greater rates. These include the increased ice flow from Greenland and Antarctica observed in 1993-2003. These rates of increase could be more or less, because the science is uncertain.

The predictions given in the 2007 IPCC report do not consider the possibility of increased rates of greenhouse gas emissions from the oceans or the permafrost, or increases in the rates of ice loss from the polar ice caps; that science remained uncertain. (Melting ice stays at 0° C until it is all gone. And with continued heating, you know what happens to the water temperature after that!) The IPCC results are consequently rather conservative. On the other hand, there may be changes resulting from international emission agreements.

These possibilities are considered, but not with a high degree of probability. In any case, the effects of climate change currently being observed cannot be easily reversed. The very long time scale for storing the excess atmospheric carbon dioxide in limestone or the deep ocean precludes a return to previous levels in our lifetime.

It has been suggested that a goal of 2° C (3.6 °F) with CO_2 of 450 ppm be adopted for managing climate change. This is considered to be a possible target for technological solutions although not a guarantee that we will avoid a tipping point caused by large CO_2 emissions from the ocean or permafrost and significant increases of melting of the polar ice caps. But implementation of such a goal continues to be in conflict with economic and political interests of many nations.

Meanwhile, climate changes demonstrated by Arctic sea ice and many glaciers indicate an acceleration of warming beyond conservative estimates. Now there is increasing concern that the feedback of reduced reflection from Arctic ice will increase the radiation imbalance and trigger early greenhouse gas emissions from the melting permafrost, causing early collapse of the Greenland ice sheet. The potential approach of these tipping points has suggested to some that a goal of 350 ppm of CO_2 will be necessary for stable climate conditions in the distant future. We are already well above that level, and approach to that goal can only be accomplished by a long and committed effort.

SECTION B

VI. FUTURE CLIMATE CHANGES

The observed temperature record and other supporting evidence of global warming is the basis for theoretical predictions of future temperature increases. An understanding of anticipated local conditions is necessarily less precise, but some generalizations can be made. Possible scenarios for the degree-by-degree increases have been presented in the National Geographic video "+6 Degrees", in the ABC program "Earth 2100", and the BBC/Discovery video "Hot Planet". Some background in the physical science fundamentals will be useful and is discussed in the following:

Weather is what we see changing every day. Climate is the average behavior over a long period. That is what we have adapted to in our lives. A single windstorm or flood doesn't make the climate, but a changing pattern of these events is a climate change. Much of the weather occurs because the earth is heated unevenly by the sun. The solar radiation strikes the surface in the tropics at near normal incidence; that is the sun's rays are perpendicular to the surface. At higher latitudes, the rays are more nearly grazing; the energy is spread over a greater area and the heating is less. Consequently, the earth's oceans and atmosphere must act to transport heat energy from the tropics to higher latitudes. There is the additional complication of the seasons; the earth's axis of rotation tilts back and forth from the sun as the earth moves in its orbit about the sun.

When warm air in the tropics moves to the polar regions the temperature decreases and air molecules lose as much as 20% of their heat to the earth's surface. On the other hand, energy stored in water vapor in the tropics is almost completely released in condensation in the cold regions. Over 5 times as much energy is stored in a given amount of water vapor as would be required to heat the same amount of water from freezing to boiling. The energy is stored slowly by evaporation in the tropics. It may be released suddenly with condensation in a storm. With global warming, there will be an increased role of heat transport from the tropics. More violent storms may be expected.

Occasional episodes of extreme weather are likely to be written off by many as false indicators of climate change caused by global warming. This will be especially true for remote storms that the media note as brief sound bites sandwiched between political disasters or celebrity scandals. Perhaps a statistical analysis of the frequency and severity of global weather events will establish the validity of climate change in the minds of climate experts, but not necessarily in the media, the policy makers, or the general public. On the other hand, the landfall of a category 5 hurricane on Manhattan or an uncontrollable wildfire enveloping Los Angeles just might stimulate some serious attention.

Decades to centuries in the future, the climate will depend greatly on humanity's response to the current pending danger. Current procrastination and business as usual will extend and increase the global warming rate into the immediate and foreseeable future. Inconvenient adaptations to occasional violent storms, coastal erosion from sea level rise, as well as disruption of water and food resources will be increasing problems. The tipping points for accelerating CH_4 and CO_2 emissions from natural reservoirs, and responding sea level rise may loom dangerously close even for the suggested 2 °C (3.6 °F) rise with 450 ppm CO_2. Mother Nature may in fact require frantic responses, including extravagant but uncertain measures of climate engineering such as satellite mirrors and injection of potentially dangerous stratospheric aerosols.

An intelligent assessment of the mistakes of cheap energy at the expense of approach to climate disaster may drive societies to take economic steps to clean non-fossil fuel energy. Even then, political and economic inertia is likely to slow corrections for a generation or so. This opens up the distinct possibility that difficult and expensive adaptations will be required for repair of coastal infrastructure, food production and distribution, and population relocation. Hopefully, the potential danger will become evident to humanity's leadership and stimulate rapid reduction of greenhouse gas emissions and initiate a long-term policy to reduce CO_2 levels to something like the proposed 350 ppm for a stable climate.

VIII. ADAPTATIONS

The IPCC report for a likely range of greenhouse gas emissions predicts a temperature increase of 0.2 °C (0.36 °F) per decade for the next two decades. So 20 years from now we may expect an average global temperature increase of 0.72 °F. Of course, that temperature increase will not be the same for all locations and precipitation may be quite variable. Warming is expected to be greatest over land with increased precipitation in high latitudes and decreases in most subtropical land regions.

There are numerous reports by government and university

scientists of the steady rise of tropical ocean temperatures and the increasing reservoir of heat energy in atmospheric water vapor. These scientific measurements and the data on increasing carbon dioxide in the long series of observations from Mauna Loa are continually added to the basic information of the atmospheric science in the atmospheric models. These reports support the recognition of the natural transport of increased heat from the tropics to higher latitudes. The conclusions in the government report of likely increase of violent weather is briefly recognized in the media; the reader will find that terrorist activities, titillating gossip about celebrities, and gun violence continue to occupy the front pages and prime time news. The not-so-well publicized and less spectacular scientific reports remain largely ignored. A prompt need for precautions against future climate developments is being neglected.

Nevertheless, the natural laws of our planet will respond to mankind's increase of the atmospheric greenhouse gases. Climate change is already underway and we must adapt. For many people, the changes may be modest and adaptation may be trivial. The early arrival of springtime temperatures after uncomfortable winters may generally be welcomed by young and old. Climate change is good! Put away the winter clothes and snow shovels. But as global warming occurs, the hummingbirds could be arriving before (or after) the flowers bloom and the mountain pine bark beetles could produce multiple generations per year to decimate the forests. Seasonal timing and temperatures will be out of kilter. Humans are greatly adaptable, but pikas and polar bears are already in trouble. So it will be survival of the fittest: we may lose a few creatures, but many humans will not notice such changes in their environment.

SECTION B

It will be important for individuals and societies to plan for modest adaptations to some of these changes. These adaptations will generally not be without cost. And there are those environmentalists with extremes of green attitudes who will prematurely identify certain of these as causes and preludes to climate disasters. Still, the supermarket manager will of course explain the price increases of red meat as the natural response to a "market economy". And yes, one may find that last season's drought on the Great Plains decreased the feed for beef cattle and the middleman response was prompt and permanent. There may also be future episodes of huge thunderstorms that could decimate the Iowa corn crop with a six-inch layer of marble size hail. This could be the result of climate change or just one of those uncertainties in the life of a farmer. But it might easily require a refinancing of the bank loan for the farmer's $300,000 tractor. Such weather extremes are well-known sporadic events that farmers and grocery shoppers have experienced for years. And your retired rich uncle living in the assisted living facility will remind all who will listen that the storms of '36 were almost as bad.

Let us consider some possible adaptations that could occur. Changes in temperature and precipitation could require different agriculture practices to produce suitable crops. These in turn would require changes in food prices and availability. Wheat farmers in Kansas may abandon their desiccated farms for virgin soils of climate-warmed Manitoba. Vegetable farmers may move their operations to Alaska. Householders may no longer find their favorite local zucchinis in the supermarket. Will Matanuska Valley produce taste the same? And will the residents of northern latitudes welcome these competitors in their new environment?

Residents of the urban-forest areas may find that the climate changes from global warming have introduced a plague of pine bark beetles and caused wildfires in the drought stricken forests. Mountain property values may plummet and owners be forced to migrate to subdivisions on abandoned farmland.

Residents of the Denver Cherry Creek neighborhood could find that their water bills require budget readjustments. The increasing costs of deep wells into the Ogallala aquifer may have inspired farmers to indulge in increased irrigation with mountain snow melt supplies, but management of these reservoirs will become complicated by the unpredictable changes in early season temperatures. An increasing popular solution for city dwellers might be a xeriscape modification of the front lawn—but probably after a bitter overhaul of homeowner agreements. And there may be a need to remove that beautiful silver maple shade tree damaged in an early spring windstorm. This could be rationalized by the increased living room sunlight but the resultant increased dependence on the air conditioner could be complicated by late summer brownouts due to more frequent double-digit afternoon temperatures.

And what might be the necessary adaptations to sea level rise? If we extrapolate the recent observation of 3.1 mm/yr rise for 20 years, we find a 6.2 cm (2.4 in) increase. This would hardly seem to be a problem. But again this is a global average; it will not be uniform. Much of this rise is due to expansion of the warm water, and ocean currents will make this a special problem for some locations. For example, the reverse of the El Nino – it's called the La Nina - will occasionally move the warm Pacific wa-

SECTION B

ter back to threaten many of the low-lying Pacific atolls. Natives will have no option other than to abandon their homes and hope that some other nation will find space for them to renew their lives. The rising seas are already causing this migration problem for Tuvalu in the Pacific and the Maldives in the Indian Ocean. There is the view that the developed countries are primarily responsible for the increased greenhouse gases and must acknowledge the substantial debt to those societies at the greatest immediate risk of climate change.

But the story of migration of Tuvalu natives from their island homes to New Zealand is no longer worthy competition in our country's news to the distress in the damage from category 3 hurricanes on the U.S. Gulf Coast, and possibly more frequent future storms. One may also expect increased Carolina to Florida beach erosion caused by more frequent Atlantic storms with a long fetch of persistent northeast winds, aggravated by continuing sea level rise. Real estate values of condominiums from those coastal retirement communities will have diminished disproportional to those in the Appalachians. Adaptation to this nuisance could be delayed if government funds for beach replenishment lag the sand erosion and sea wall damage. One should also anticipate assessments of increasing condo upkeep and roof repair which could inspire increasing numbers of retirees to list their property with a broker and migrate to the sun country of New Mexico and Arizona -- where they may vie for space and for water with migrants from other coastal nations. A few gullible New Jersey seniors, ignorant of the rising sea levels, will probably continue to invest their life savings in beach property. Others may simply skip adaptations and plan to enjoy the balmy breezes for their golden years: their offspring's inheritance and society's future climate problems are absent from their thoughts.

VIII. SOLUTIONS

Climate change is real! The major cause for recent climate change is the greenhouse effect of the heat trapping absorption of earth's heat radiation by molecules of water, carbon dioxide, methane, nitrous oxide and others. The balance has supported life as we know it, with earth surface temperature at least 20 degrees Celsius (36 degrees Fahrenheit) above that expected for an atmosphere-free planet. Recent observations of global warming indicate that the increase of these greenhouse gases, particularly CO_2, is presently the principal driver of the earth's climate change. It is the future increases of these greenhouse gases that will continue to affect life on our planet. The effects are not globally uniform; in the future, some will be found good, others disastrous. Adaptation is possible for many but not all human societies. Some plants and animals will adapt, others will disappear.

Recognition of global warming and its human causes is sometimes rationalized and sometimes welcomed by some individuals. Their logic is frequently flawed by narrow economic considerations and ignorance of the science. For example, they may argue that more CO_2 should increase plant growth and benefit the starving poor of distant lands. They recommend that we should respond by clearing of forests and increasing irrigation and fertilizer applications. But there will be consequences of precipitation changes from the changing climate: crops do not do well in the desert! Finally, the science warns us that increased CO_2 and N_2O

SECTION B

greenhouse gases from increased agriculture would accelerate global warming and disruptive climate change.

It is clearly recognized that human activity is responsible for much of the greenhouse gas increase. Some is due to agricultural practices such as fertilizer use on plants and food production by raising animals. This production of nitrous oxide and methane yields greenhouse gases that, molecule for molecule, are exceptionally strong absorbers of earth's heat but have a relatively short lifetime in the atmosphere. Decreases in their emission would result in a relatively prompt decrease in global warming, but at a great expense in food production. A larger fraction of the temperature increase is due to burning of large amounts of fossil fuels for energy production. Human societies have enjoyed the use of cheap energy but have avoided the costs of current and very long-term climate change. It is this energy production and use factor that requires a new mix of technologies and human behaviors.

Human societies have considerable control over the emission of CO_2 from burning fossil fuel and producing cement from limestone. Humans also have some influence on increasing the carbon storage in forests by slowing deforestation or increasing conservation efforts. It has been suggested that we might fertilize the ocean with an iron compound to increase the rate of carbon ultimately stored in the deep ocean by biological activity. There is even the possibility of future technologic removal of CO_2 from the atmospheric reservoir. Details of some solutions need careful study and consideration.

An obvious direct solution is to minimize our energy use and to produce that necessary energy from alternate sources rather than from fossil fuels. Earth receives thousands of times as much energy from the sun as that generated for all the electric power on earth. It follows that harnessing a small fraction of the solar radiation through use of solar voltaic or solar hot water systems would satisfy our energy needs for electricity or heat with no production of CO_2 greenhouse gas. Photovoltaic technology produces electric energy directly. Solar thermal processes absorb solar radiation to generate heat. Modest systems are used for heating homes and other buildings; more elaborate systems produce high temperature for use in conventional steam-powered electric generation. Cloud-free areas with long periods of overhead sun are making increased use of this energy source.

Weather systems that generate wind are an example of indirect solar energy. Windmills have been in use for pumping water and grinding grain for centuries. Larger, more efficient turbines for generation of electricity are becoming available in regions with dependable winds. Again, there is no production of CO_2 greenhouse gas from this source of electrical power.

The development and investment in these technologies has lagged behind the acknowledged need for alternative energy because of market uncertainties from government policies. There is a related problem in construction of transmission facilities to deliver this energy to high-use cities. Some environmental problems concerning water and wildlife affected by introducing new human elements into their habitats will also require attention.

Burning hydrogen gas is another non-CO_2 pollution process for energy production. The combustion process produces only H_2O. There is an obvious challenge in generating and distributing the hydrogen. Another alternative energy source is biofuel. Production of heat energy by burning wood for homes and cooking has a long history. While the combustion product is largely CO_2, the fuel is produced and replaced by photosynthesis of tree growth, removing CO_2 from the atmosphere. Unfortunately, our planet's forests have been greatly depleted by deforestation for timber and agriculture. These practices could be reversed and regulated. We could also learn to manage our forests to improve carbon storage. Healthy trees approaching maturity are the most efficient of nature's plants in converting atmospheric CO_2 to carbon in their wood. We may remove trees suitable for fuel production but we should take care to maintain a healthy forest for carbon storage. Our practice of thinning of lodgepole forests for wildfire protection should continue, for example, but mature trees should be preserved for ongoing carbon storage. Studies have shown that the mature coastal redwoods of California produce quality wood at an increasing rate to an age of 1600 years. Timber production for construction or eventually for fuel that maintains these old growth forests can be an excellent means for storing atmospheric carbon. An equilibrium of tree growth and energy production by humans can thus be carbon neutral, but only if we control forest fires, deforestation for timber or agriculture, and other demands of uncontrolled population growth.

Energy for auto, air, and rail transportation, and to some extent home heating, relies heavily on fossil fuel oil. Burning this fossil fuel is a large contributor to the CO_2 greenhouse gas responsible for global warming. The environmental climate change cost of burning oil has finally been recognized, yet it is independence from foreign oil that is usually cited as the need to develop alternate fuels. The alternative energy production from biofuels has various problems. For example, corn and palm oil are major food sources in many parts of the world: the availability and cost of food from these sources is being affected by these competitive demands of energy production. Biofuels are nevertheless important as a partial solution to the global warming threat since they offer the advantage of an approach to carbon neutrality; the CO_2 emission may be balanced by plant growth of the biologic energy source. Finally though, the production and environmental costs of this energy alternative must be justified with the extent of approach to carbon neutrality.

Nuclear fission presently generates a small fraction of the world's electric power. A limiting disadvantage is the presence of radioactive decay products and the threat of nuclear proliferation for terrorist bomb production. Fission of the nuclei of heavy elements like uranium yields a pair of nuclei whose total mass is slightly less than that of the parent. This mass loss produces large amounts of energy following the Einstein relationship $E = mc^2$. A few neutrons are also produced that can initiate further fission processes in a chain reaction. Neutrons slowed by collisions with stable light elements are useful in generating power in controlled reactions. Unfortunately the daughter nuclei are radioactive, emitting damaging electrons and gamma rays, and managing these waste products is a problem. Highly enriched fissionable nuclei have also been used to produce tremendous destruction in uncontrolled fission explosions, posing a worrisome security threat. However, France has an excellent safety record and generates a large fraction of its power with nuclear

SECTION B

fission. While the expansion of nuclear fission power may be considered a viable non-CO_2 power source for the future, significant delays in the approval and construction of new fission reactors are a concern for prompt correction to global warming and disruptive climate change.

Nuclear fusion reactions also generate large amounts of energy from $E = mc^2$ when nuclei of light elements (heavy hydrogen (deuterium), for example) are combined to make a stable nucleus of slightly less total mass (helium). Combining these charged particles in collisions requires that the fuel be heated initially to very high temperatures. This high temperature plasma of charged particles must be confined, perhaps by magnetic fields. (The sun produces energy by fusion; the high temperature plasma is confined by gravity.) Research on the heating and confinement processes has been underway for 50 years; a pilot reactor is currently being constructed in France with international support. It should be noted that the product nuclei are not radioactive; there is no waste disposal problem as there is with fission. The hydrogen bomb is an uncontrolled explosion, but this can only occur with extremely sudden heating of the dense source material. The controlled power generation through initial production and confinement of the hot plasma is difficult. By the same token it is a safety factor; a failure to control the hot gas would automatically halt the fusion processes. Fortunately the source of light nuclei is virtually unlimited. Fusion power has the potential to become the ultimate source of large amounts of our energy in the somewhat distant future.

Coal and natural gas as fuels for electric generation is a well-developed technology that is currently the most economical energy source. Unfortunately, it is also a major source of the CO_2 greenhouse gas. It is finally being recognized that this cheap energy exists because the cost of the environmental climate change pollution has never been charged to the consumer. (This is not trivial. Imagine the cost of cooling the oceans to reduce the sea level rise of the past century.) The costs of electric power from alternative sources, such as photovoltaic, solar, or wind, are presently somewhat greater than the costs from fossil fuel sources, but there are no direct environmental climate change costs.

Some relief from the danger in increasing CO_2 greenhouse gas from fossil fuel energy production could be achieved with energy conservation by consumers. Energy efficiency in buildings-government, commercial, and private-could be effective through careful usage, improved insulation, and other design features. Such reduced energy use has the obvious dual advantage of reduced direct cost as well as reduction of CO_2 emissions.

Technologic solutions to correct the economic advantages of polluting fossil fuel production have been slow in the United States in recent years. Research advances and mass production of the alternative sources will be at least a partial economic correction to make alternative power less costly. Research and development of techniques for CO_2 capture and sequestration from fossil fuel facilities is an additional production cost for fossil fuel sources, but it would also be a potential solution to permit continued use of fossil fuel sources without CO_2 pollution. Although a governing policy to diminish CO_2 emissions would initially cost more than ignoring the global warming problem, it would narrow the economic gap with

alternative sources and encourage industries that would ultimately permit human societies to cope with climate change.

Policies to minimize global warming received economic support when a number of developed countries like Germany,
France, Denmark, and Japan sought to correct their limita-
tions on energy from fossil fuels with long-term government
support of alternative energy sources. China has become an
extremely high greenhouse gas contributor due to its technological development supported by very large reserves of coal.
However, with knowledge of the environmental need to limit
greenhouse gases in the very near future, the Chinese government has invested heavily in support of alternative energy
industries that may ultimately establish its world leadership in
solar and wind technologic equipment. In the United States,
individual and corporate efforts to utilize alternative energy
sources presently rely largely on European hardware. Failure
of our governing bodies to recognize the economic policy
need to support alternative energy industries in the face of
disruptive climate change may logically result in abandon-
ment of leadership in economic as well as a moral role in
future world affairs.

Economic concerns will necessarily be a factor in the sched-
ule of shift to alternative energy sources. A reliance on
economic control of energy production might ultimately lead
to more competitive use of alternative energy, but there is the
danger that the time delay will be too great to avoid exten-
sive disruptive climate change. This delay could be alleviated
by recognition of the historical neglect to account for the
climate change cost of the use of fossil fuel for energy production. An obvious correction is to legislate a carbon tax or

a cap and trade policy. For those who advocate procrastination in this respect, one should think of this as an insurance policy. Homeowners generally pay for fire insurance on their houses, even though the threat does not seem real or imminent. Disruptive climate change is at least as real and may be imminent.

IX. LESSONS FROM THE STRATOSPHERIC OZONE PROBLEM

The stratospheric ozone problem is a history lesson that is appropriate to remember as we face the problem of global warming. Ozone is a naturally occurring molecule consisting of three oxygen atoms; the oxygen we breathe has only two. The earth has a layer of ozone in the stratosphere that absorbs most of the ultraviolet reaching earth from the sun. Possible damage to the ozone layer by man-made chemicals was well-understood and recognized as a global problem because people everywhere would be at risk for increased sunburn and the possibility of developing skin cancer if there were less ozone. We quickly learned to use sunscreen, and with unprecedented international cooperation agreed to stop production of the harmful chemicals and utilize substitute compounds.

Let us look in more detail at that problem. The sunlight controls the chemistry of the atmosphere; it makes and destroys ozone every day maintaining a balanced amount that we have learned to live with. But about 25 years ago chemists synthesized a molecule that made air conditioning nearly universal for homes and cars, and was useful as a propellant in spray cans. These chemicals, called chlorofluorocarbons

(CFCs), did not harm the environment around us—the air, the water, the trees, the birds, the animals—but some research scientists warned that the large increasing reservoir of this compound in the atmosphere had the potential to thin the stratospheric ozone layer resulting in more ultraviolet reaching the surface of the earth. So we put all the known chemistry in the computer models of the atmosphere, and yes, there was a little problem that we would have to think about, but no hurry. But nature had a surprise for us. Atmospheric observers in Antarctica observed a hole in the ozone layer in the springtime. There were chemical reactions on the polar stratospheric clouds that released chlorine compounds that destroyed all the ozone for a couple of months every spring until the ice clouds disappeared. We had failed to include those chemical reactions on polar stratospheric clouds in the computer model. The ozone problem was a lot more serious than originally thought.

Fortunately, our mistake in chemistry was easy to correct. Substitute compounds were already available. We fixed our mistake by quickly making international agreements to stop manufacturing and using the offending compounds and we developed a substitute for use in air conditioning. The problem is not expected to get worse, but Nature is taking her time about repairing the damage. We still have a reservoir of man-made chlorine compounds with long lifespans in the atmosphere. Those molecules will continue to diffuse into the stratosphere and destroy ozone. The ozone holes will not disappear for 50-75 years.

Climate change with global warming is not so simple. The danger is not the same for everyone, but correcting it involves major changes in energy production for everyone. CO_2 will remain in the atmosphere even longer than the CFCs. If we procrastinate in fixing the problem, the situation will not only get rapidly worse but also the time required to repair the damage will be much longer.

The stratospheric ozone problem occurred before middle school students were born and most other folks have forgotten about it. But there are lessons to be learned:

Lesson I: If we don't completely understand the problem, nature may surprise us. We make predictions based on what we know. Because we did not know that there were ice clouds over Antarctica with different chemistry than what we understood and incorporated into our computer models, the ozone holes were a complete surprise. With climate change, there are also processes that are not well understood—like the movement of the Greenland glaciers or the methane and carbon dioxide release from the oceans and the melting permafrost—or other things we have not even imagined yet. So there may be some unpleasant surprises that are not in our predictions about future effects.

Lesson II: The science is different. Ozone was chemistry, but carbon dioxide is infrared heat radiation. The CFC and CO_2 polluting molecules do have one thing in common: once we put them in the atmosphere we can't easily get them out. Those man-made reservoirs don't disappear quickly. Yesterday's mistake will continue to affect the earth for a long time. The ozone holes won't heal for 50-75 years, and the CO_2 molecules that we add to the atmosphere today will continue absorbing and trapping the earth's heat for 100 years or longer. We need to correct our environmental mistakes now; nature's laws will not kindly or quickly move us back to square one.

Lesson III: To solve the ozone problem we cooperated promptly to make international agreements to stop manufacturing and using CFCs. The mistake was corrected and nature will eventually fix it. It could have been worse. Slowing or stopping global warming will be much more difficult—we can't just stop producing energy or growing food. It is a global problem requiring global cooperation: treaties, laws, rewards, whatever it takes to reduce the current rate of production of greenhouse gases. If we delay, the problem will just keep getting worse and more expensive to fix.

X. CONCLUSIONS

The greenhouse effect of atmospheric CO_2 is firmly established and the data from Mauna Loa clearly show the 30% increase of this pollutant in the atmosphere. There is also no doubt that this increase is largely due to human activities, especially the increased burning of fossil fuels for energy production. The observation of melting of glaciers and Arctic ice and warming of the oceans indicate very large increases of energy on the Earth's surface that can only be attributed to heat trapping by the greenhouse effect.

Climate change effects from the increased trapped energy are complex. The planet's transport of heat from tropics to polar regions will be affected in a variety of ways in weather events and ocean currents. Temperatures should generally reflect the energy increase but there may be geographic variations due to the above transport mechanisms. These environmental changes will require adaptations of various life forms, including humans. The reservoirs of carbon storage in soils and oceans are potential sources of large and possibly sudden increases of atmospheric CO_2 due to increasing temperature at some uncertain future time, which would cause additional considerable disruptive climate changes. Intelligent humans in earth's societies should recognize the validity of natural environmental laws that respond to pollution by greenhouse gases, and proceed with corrections in energy and food production before the planet reaches a tipping point to severe disruptions in climate.

We are in the habit of thinking of short-term results, like next quarter's profits or fixing yesterday's problems. But dollars spent on solar panels and wind turbines now will give us a return of free energy in the not too distant future. Can't spend money on that just now? But if you have money in the bank, it's because you had the benefit of cheap energy without paying for the pollution that's now threatening the planet.

The cathedral builders of the past were dedicated to a beautiful life for future generations. Yes, we invest in our children's education; should we not also invest in their future environment?

References

CNA. 2009. Powering America's Defense: Energy and the Risks to National Security. Available at www.cna.org/nationalsecurity/energy/

Diaz, Henry F., and Richard J. Murnane, eds. 2008. Climate Extremes and Society. Cambridge University Press.

Goddard Institute for Space Studies. 2009. GISS Surface Temperature Analysis. Analysis Graphs and Plots. Available at http://data.giss.nasa.gov/gistemp/graphs/

Hansen, James. 2009. Storms of My Grandchildren. Bloomsbury USA.

Harrould-Kolieb, Ellycia, and Jacqueline Savitz. 2009. Acid Test: Can We Save Our Oceans from CO_2? Available at www.oceana.org/north-america/publications/reports/

Houghton, John. 2004. Global Warming, The Complete Briefing. Cambridge University Press.

Intergovernmental Panel on Climate Change (IPCC). 2007. Climate Change 2007: The Physical Science Basis. Available at www.ipcc.ch/publications_and_data/publications_and_data_reports.htm

Karl, Thomas R., Jerry M. Melillo, and Thomas C. Peterson. 2009. Global Climate Change Impacts in the United States: a state of knowledge report from the U.S. Global Change Research Program. Cambridge University Press. Available at www.globalchange.gov/publications/reports/scientific-assessments/us-impacts

National Research Council, National Academies of Science. 2008. Ecological Impacts of Climate Change. Available at www.nap.edu/catalog/12491.html

Scientific Expert Group on Climate Change. 2007. Confronting Climate Change: Avoiding the Unmanageable and Managing the Unavoidable. Research Triangle Park, N.C., and Washington, D.C.: Report for Sigma Xi and the United Nations Foundation. Available at www.unfoundation.org/files/pdf/2007/SEG_Report.pdf

Schneider, Stephen H. 2009. Science As A Contact Sport: Inside the Battle to Save the Earth's Climate. National Geographic. Second Edition June 2009

Suggested Additional Reading

Cook, Alex. 2007. The Greenhouse Effect, a Legacy: A Novel of Living With Climate Change and A Scientist's Brief on Global Warming. Dog Ear Publishing.

Craven, Greg. 2009. What's the worst that could happen?: A Rational Response to the Climate Change Debate. Perigee Trade.

Flannery, Tim. 2005. The Weather Makers. Atlantic Monthly Press.

Gore, Al. 2010 "We Can't Wish Away Climate Change". Op-Ed piece, New York Times, 28 February 2010. Available at www.nytimes. com/2010/02/28/opinion/28gore.html?scp=2&sq=al%20 gore&st=cse

Gore, Al. 2009. Our Choice: How We Can Solve the Climate Crisis. Rodale Books.

Gore, Al. 2006. An Inconvenient Truth, The Planetary Emergency of Global Warming and What We Can Do About It. Rodale Books.

Kammen, Daniel M. 2006. Energy's Future Beyond Carbon. Scientific American Special Issue, September 2006, Vol 295, #3. Available at http://rael.berkeley.edu/sciam0906

Kolbert, Elizabeth. 2006. Field Notes from a Catastrophe: Man, Nature, and Climate Change. Bloomsbury USA.

Krupp, Fred, and Miriam Horn. 2008. Earth: The Sequel. W. W. Norton and Company.

Linden, Eugene. 2006. The Winds of Change. Simon and Schuster.

Lovelock, James. 2006. The Revenge Of Gaia: Earth's Climate Crisis and the Fate of Humanity. Basic Books.

Lynas, Mark. 2004. High Tides. London: Flamingo

National Geographic Society. 2009. The Carbon Bathtub. National Geographic 216(6):26-29. Also http://ngm.nationalgeographic.com/big-idea/05/carbon-bath

Ochoa, George, Jennifer Hoffman, and Tina Tin. 2005. Climate. Rodale Books.

The Prince's Rainforest Project. http://www.rainforestsos.org/content/home/ March 2009

Videos

An Inconvenient Truth. Al Gore

The Dimming of the Sun. Nova

Green, the New Red, White, and Blue. Tom Friedman, Discovery Channel

Hot Planet. BBC and Discovery channel

How It All Ends. Greg Craven.Available at www.youtube.com/ wonderingmind42

The Power of the Sun. Walter Kohn and Alan Heeger, University of California

Richard Alley Dances to Explain Ice Ages, CO_2 and Global Warming. Available at www.youtube.com/watch?v=-NQPolcYoIc

Saved by the Sun. Nova

Six Degrees. Mark Lynas, National Geographic

Appendix

"A Contribution to the Lynn Journal" from
The Greenhouse Effect- A Legacy,
by Alex Cook

It is with considerable humility that I take this opportunity to communicate my analysis of the subject of global climate change to you. It is an obligation that comes with the inheritance of proper genes sufficient to obtain a scientific degree and that modicum of wisdom about life in general that comes with age. What I have to tell you is not new information.

There are facts about the changing climate that are incontrovertible. These are not in the class of political opinions; they are part of the present body of generally accepted science. Let me first summarize those that are not subject to the whim of politics, religion, or talk shows. Observed surface temperatures have shown a modest upward trend for the past century, and the rate is increasing. The world's glaciers are melting at an increasing rate, the Arctic ice is decreasing, and the permafrost is melting. Sea levels are slowly rising. All are consistent with global warming.

How could this happen? Well, you see the earth is actively supplied with energy from the sun. By virtue of the sun's extremely high surface temperature, its maximum radiation intensity is in the visible region of the spectrum. The earth, in a near circular orbit at 93 million miles from the sun, receives a fraction of this energy and loses heat only by radiation. At the earth's lower temperature, the outgoing radiation has maximum intensity beyond the visible, in the infrared. This balance of incoming solar radiation and the outgoing heat radiation determines the earth's surface temperature. But if there were no atmosphere, the earth would be expected to have an average surface temperature not quite warm enough to encourage life as we know it.

Actually, the outgoing infrared is partially absorbed by certain molecules in the atmosphere, especially water vapor; the heat radiation resulting from the processes of absorption and re-radiation is trapped between this atmospheric blanket and the earth's surface. The planet's temperature is substantially warmer. We call this the Greenhouse Effect, in analogy to the operation of structures for protection and growth of flowers and vegetables. Life on earth has benefited from the substantial water vapor in the atmosphere that contributes to the greenhouse effect. This infrared absorption has yielded an average surface temperature that is about 38 degrees Fahrenheit above that for an earth with a complete absence of the greenhouse effect.

Laboratory measurements demonstrate that carbon dioxide molecules, produced in the burning of carbon containing materials, are another strong absorber of infrared radiation in the spectral region of maximum outgoing radiation from the earth. It is thus one of the most important greenhouse gases that constitute the infrared blanket.

Human populations are increasing and technology shows rapid advances. These are fueled by exploitation of our planet's resources for food, shelter—and for energy. There is a measured 1/3 increase in the carbon dioxide greenhouse gas, clearly related to the combustion of carbon containing fuels. It is this enhanced greenhouse effect that can alter the earth's radiation balance.

Those are the facts.

Scientific models of climate, consistent with the above facts, use high-speed computers to do atmospheric model calculations to be verified with present climate and to predict future changes. These yield a consensus that increasing atmospheric concentrations of carbon dioxide are presently enhancing the global warming, and that continuing increases of this atmospheric constituent will yield further temperature increases. These predictions of the future based on the facts of science can be made with considerable confidence.

Humans have become an influential part of Nature, demonstrated in part by the facts of present climate change. We have increased the atmospheric carbon dioxide—and its lifetime in the environment is on the order of one hundred years. Consequently, we are already committed to global warming; it is now a question of how fast and how bad. Our immediate future and the extended future of our offspring depend on our response.

Already, global warming has led to disaster in some regions, such as the sub-Sahara, where failure of natural resources has led to genocide. On the other hand, a very large part of the world's population presently, or will soon, enjoy a standard of living supported by heavy energy consumption, fueled by coal and oil. This is the principal contribution to further global warming.

It has been said that climate change with global warming is an inconvenience. For those whose profits are dependent on energy production from fossil fuels, it is a minor embarrassment; major changes in their source of wealth may become necessary. For most of us, who must adapt to an increased rate of change in our environment, there will be expensive nuisances. But for those humans and other creatures whose very existence is attuned to their present environment, an inconvenience is a major understatement.

How does one cope? Denial is a favored technique for many. For those with little inclination for careful thought, this may be supported by a misguided faith that Mother Nature is an inviolate entity. She operates independently of humans. Hopefully she will be kind. Some, who are scientifically ignorant, will scoff at the record of temperature measurements and atmospheric model predictions. Others, with selfish interests in power or profit, will trust that disaster can be postponed beyond their time. Their denial is based on false interpretations of the scientific observations. Results can be debunked or censored for consumption by the media and gullible public. Or a corrective response can be delayed simply by a 'need for more research', a statement as sacred as motherhood.

But none of these will alter the planet's response to the natural laws of the universe. The greenhouse effect will continue to operate. Humanity's increased gluttony for cheap energy with related production of greenhouse gases makes us an influential part of Nature, with inevitable consequences.

In general, our U.S. population has been blessed with copious natural resources; these have been exploited to economic advantage. A large segment of our society has become arrogant in the expectation that we deserve a high standard of living, irrespective of our role in Nature or the aspirations of other societies. We live in large comfortable houses and drive our choice of several automobiles. We can afford modern devices for communication and entertainment. Our successful leaders are CEOs of energy or automobile industries; the rest of us follow. The policies of our government are driven by economic success. The environmental attitude borders on denial of global change or human influence. The developing nations follow our example. The poor nations have no choice; they have relatively little influence in global matters.

We may extrapolate this policy into the near future. A consequence is clearly the increase the greenhouse gas concentrations; global warming will inevitably accelerate and become more intractable in the future. Our nation will have sacrificed its responsible leadership. We will have fewer friends and more enemies in other societies.

Suppose we choose to disregard the threat; we may distort the interpretation in order to favor personal or other special interests. Business as usual. We will rely on the power of our economy and technologic advances to cope with any imagined future problem. Our energy-dependent society might hope for technological weapons to combat the forces of nature, or to facilitate future adaptation. In the meantime, it is tempting to continue a policy of emphasis on controlling the sources of fossil energy to fuel a growing economy.

But, global warming will not make exceptions for the individual citizens of this nation. There will be increasing events of weather discomfort. Environmental degradation will become increasingly apparent with biological failures in plant and animal communities. If we stay the course, we must crank up the air conditioning, repair and guard against beach erosion, fight increasing forest fires and proliferation of damaging insects, construct elaborate systems to ensure water for drought stricken farms and cities, and repair the ravages of more violent storms.

We may temporarily succeed in these endeavors by virtue of our wealth and energy use. Our wealthy nation can make adaptations to climate change. Technology is expected to develop means to sequester atmospheric carbon dioxide without sacrificing energy production. We have the capability to construct water distribution systems to revive drought regions, desalinization plants for cities, increased fire suppression for wildfires in the dry forests, coastal armoring to combat beach erosion, and relocation of coastal populations.

Each of the above requires considerable economic support. Shortfalls must be taken care of by increased taxes or increased debt. We, the polluter, will be forced to pay; the war will be costly and must be financed by taxes or increased debt to other nations. Taxes promise to increase the personal pain already inflicted by climate discomforts. The mortgages of our citizens and the nation will become due. Other nations will have similar problems; we may expect general economic difficulties. And we have worsened the problem of global change. The continuing increase in energy consumption and resultant CO_2 increase will accelerate the approach to climate disaster. Ultimately it will become apparent that we have lost the battle with the natural laws of the Universe.

In contrast to a 'business as usual' policy, there exists a small segment of our educated population who possess some factual knowledge of the environment and tend to think for themselves. They occasionally abandon sitcoms to become familiar with global events described on the Discovery Channel or in the literature designed for public analysis. While the future is never entirely predictable, we might take advantage of the 'precautionary principle'; by that, I mean that we generally endorse the concept of insurance against future financial disasters, and military buildup against potential attackers. The potential for damaging climate change is no less real. These individuals may take steps for personal energy conservation, support of alternative solar and wind energy production, and reforestation projects. These corrective actions are a positive beginning--with universal adoption, one might hope for a modest reduction in the rate of global warming.

Actually, some of our cities and states have independently adopted a more responsible attitude and corrective action with conservation and improved energy efficiency. And other enlightened nations have adopted the Kyoto protocol to reduce greenhouse gas emissions. Perhaps before human civilization begins a final descent to extinction, some constructive efforts will succeed in coping with this human-caused climate change.

The intelligent world may come to an understanding that we have a powerful common adversary in the natural laws of the universe. Cooperative actions might then occur among all civilized nations.

These would include conservation and increased efficiency in all energy consumption. Forests and oceans would be managed for increased carbon sequestration. Transportation methods would be modified. Public transportation would be emphasized, supplemented by electric powered taxis. Personal automobiles would be designed for efficiency. Air transport would be essentially limited to private matters or large conferences; ordinary business affairs could be conducted electronically. Electrical power would be reengineered with a crash program of nuclear fission reactors; nuclear warheads would be reconstituted for peaceful power. The research and development of fusion reactors would be given long-term emphasis. And finally, when electrical power is nearly CO_2 free, hydrogen will become available by electrolysis to fuel trains and buses—perhaps even automobiles; the exhausts will contain only water vapor!

But there are potential climate disasters whose time scale is uncertain. We need to expand our use of other alternative energy sources in the very near future. There are relatively small-scale methods whose lead-time should be short. Solar, wind, and biofuel technologies are being rapidly developed in other societies. Our nation should learn from their successes.

Predictions based on human intelligence, aggression, or compassion are uncertain. But faced with the prospect of future civilization disaster from continuing climate change, humans must recognize and cope with the impersonal overpowering entity of natural laws. We may call it Mother Earth, or Gaia, after the concept of a complex of positive and negative feedbacks in Nature.

However, the early interpretation of Gaia as a benevolent living organism existing to guarantee human life may be a wishful overstatement. My Gaia, the complex of natural physical laws, must ultimately be recognized as the supremely powerful deity. This Goddess is not vindictive, but ruthless nevertheless. Each thought and action by individuals, and the policies of nations must recognize and yield. Intelligent leaders must become aware of the science and assume responsibility for action. Every society, whether organized about a specific industry or resource, must recognize the ultimate authority of Gaia. Every established religion must join in common respect. An enlightened and peaceful civilization may then continue to inhabit the earth.

www.ingramcontent.com/pod-product-compliance
Lightning Source LLC
Chambersburg PA
CBHW021958170526
45157CB00003B/1051